Lecture Notes in Computer Science 16152

Founding Editors

Gerhard Goos
Juris Hartmanis

Editorial Board Members

Elisa Bertino, *Purdue University, West Lafayette, IN, USA*
Wen Gao, *Peking University, Beijing, China*
Bernhard Steffen, *TU Dortmund University, Dortmund, Germany*
Moti Yung, *Columbia University, New York, NY, USA*

The series Lecture Notes in Computer Science (LNCS), including its subseries Lecture Notes in Artificial Intelligence (LNAI) and Lecture Notes in Bioinformatics (LNBI), has established itself as a medium for the publication of new developments in computer science and information technology research, teaching, and education.

LNCS enjoys close cooperation with the computer science R & D community, the series counts many renowned academics among its volume editors and paper authors, and collaborates with prestigious societies. Its mission is to serve this international community by providing an invaluable service, mainly focused on the publication of conference and workshop proceedings and postproceedings. LNCS commenced publication in 1973.

Shunli Zhang · Yu Zhang · Santoso Wibowo ·
Liang-Jie Zhang
Editors

Big Data – BigData 2025

14th International Conference
Held as Part of the Services Conference Federation, SCF 2025
Hong Kong, China, September 27–30, 2025
Proceedings

Editors
Shunli Zhang
QingHai Institute of Technology
Xining, China

Santoso Wibowo
Central Queensland University
Queensland, QLD, Australia

Yu Zhang
Huazhong University of Science
and Technology
Wuhan, China

Liang-Jie Zhang
Shenzhen University
Shenzhen, China

ISSN 0302-9743   ISSN 1611-3349 (electronic)
Lecture Notes in Computer Science
ISBN 978-3-032-06523-0   ISBN 978-3-032-06524-7 (eBook)
https://doi.org/10.1007/978-3-032-06524-7

© The Editor(s) (if applicable) and The Author(s), under exclusive license
to Springer Nature Switzerland AG 2025

This work is subject to copyright. All rights are solely and exclusively licensed by the Publisher, whether the whole or part of the material is concerned, specifically the rights of translation, reprinting, reuse of illustrations, recitation, broadcasting, reproduction on microfilms or in any other physical way, and transmission or information storage and retrieval, electronic adaptation, computer software, or by similar or dissimilar methodology now known or hereafter developed.
The use of general descriptive names, registered names, trademarks, service marks, etc. in this publication does not imply, even in the absence of a specific statement, that such names are exempt from the relevant protective laws and regulations and therefore free for general use.
The publisher, the authors and the editors are safe to assume that the advice and information in this book are believed to be true and accurate at the date of publication. Neither the publisher nor the authors or the editors give a warranty, expressed or implied, with respect to the material contained herein or for any errors or omissions that may have been made. The publisher remains neutral with regard to jurisdictional claims in published maps and institutional affiliations.

This Springer imprint is published by the registered company Springer Nature Switzerland AG
The registered company address is: Gewerbestrasse 11, 6330 Cham, Switzerland

If disposing of this product, please recycle the paper.

Preface

The 2025 International Conference on Big Data (BigData 2025) aimed to provide an international forum to formally explore various business insights into all kinds of value-added "agents for big data". Big Data is a key enabler in exploring business insights and economics of services.

BigData 2025 was a member of the Services Conference Federation (SCF). SCF 2025 had the following 10 collocated service-oriented sister conferences: 2025 International Conference on Web Services (ICWS 2025), 2025 International Conference on Cloud Computing (CLOUD 2025), 2025 International Conference on Services Computing (SCC 2025), 2025 International Conference on Big Data (BigData 2025), 2025 International Conference on AI & Multimodal Services (AIMS 2025), 2025 International Conference on Metaverse (METAVERSE 2025), 2025 International Conference on Internet of Things (ICIOT 2025), 2025 International Conference on Cognitive Computing (ICCC 2025), 2025 International Conference on Edge Computing (EDGE 2025), and 2025 International Conference on Blockchain (ICBC 2025).

This volume presents the accepted papers of the 2025 International Conference on Big Data (BigData 2025), held in Hong Kong, China during September 27-30, 2025. For this conference, each paper was single-blind reviewed by three independent members of the International Program Committee. After carefully evaluating their originality and quality, we accepted 15 papers from 25 submissions.

We are pleased to thank the authors whose submissions and participation made this conference possible. We also want to express our thanks to the Organizing Committee and Program Committee members for their dedication in helping to organize the conference and reviewing the submissions. We owe special thanks to the keynote speakers for their impressive speeches.

Finally, we would like to thank operations team members Jing Zeng, Sheng He, Yishuang Ning, and Zhuolin Mei for their excellent work in organizing this conference. We look forward to your future great contributions as a volunteer, author, and conference participant in the fast-growing worldwide services innovations community.

August 2025

Shunli Zhang
Yu Zhang
Santoso Wibowo
Liang-Jie Zhang

Conference Sponsor – Services Society

The Services Society (S2) is a non-profit professional organization that has been created to promote worldwide research and technical collaboration in services innovations among academia and industrial professionals. Its members are volunteers from industry and academia with common interests. S2 is registered in the USA as a "501(c) organization", which means that it is an American tax-exempt nonprofit organization. S2 collaborates with other professional organizations to sponsor or co-sponsor conferences and to promote an effective services curriculum in colleges and universities. S2 initiates and promotes a "Services University" program worldwide to bridge the gap between industrial needs and university instruction.

The Services Sector accounted for 79.5% of the GDP of the USA in 2016. The Services Society has formed 5 Special Interest Groups (SIGs) to support technology- and domain-specific professional activities.

- Special Interest Group on Services Computing (SIG-SC)
- Special Interest Group on Big Data (SIG-BD)
- Special Interest Group on Cloud Computing (SIG-CLOUD)
- Special Interest Group on Artificial Intelligence (SIG-AI)
- Special Interest Group on Metaverse (SIG-Metaverse)

About the Services Conference Federation (SCF)

As the founding member of the Services Conference Federation (SCF), the first **International Conference on Web Services (ICWS)** was held in June 2003 in Las Vegas, USA. Meanwhile, the First International Conference on Web Services - Europe 2003 (ICWS-Europe 2003) was held in Germany in October 2003. ICWS-Europe 2003 was an extended event of the 2003 International Conference on Web Services (ICWS 2003) in Europe. In 2004, ICWS-Europe was changed to the European Conference on Web Services (ECOWS), which was held in Erfurt, Germany.

Sponsored by the Services Society and Springer, SCF 2018 and SCF 2019 were held successfully on June 25 – June 30, 2018, in Seattle, USA, and on June 25 – June 30, 2019, in San Diego, USA. SCF 2020 and SCF 2021 were held successfully online and in satellite sessions in Shenzhen, China. SCF 2022 and 2023 were held successfully on December 10–14, 2022 and on September 23–26, 2023, in Hawaii, USA. SCF 2024 was held successfully on November 16–19, 2024, in Bangkok, Thailand. To celebrate its 23rd birthday, SCF 2025 was held on September 27–30, 2025, in Hong Kong, China.

In the past 22 years, the ICWS community has expanded from Web engineering innovations to scientific research for the whole services industry. Service delivery platforms have been expanded to mobile platforms, the Internet of Things, cloud computing, and edge computing. The services ecosystem has gradually been enabled, value-added, and intelligence embedded through enabling technologies such as big data, artificial intelligence, and cognitive computing. In the coming years, all transactions with multiple parties involved will be transformed into blockchain and metaverse.

Based on technology trends and best practices in the field, the Services Conference Federation (SCF) will continue serving as the conference umbrella's code name for all services-related conferences. SCF 2025 defined the future of New ABCDE (AI, Blockchain, Cloud, BigData, & IOT) and entered the 5G for Services Era. **The theme of SCF 2025 was Services Agent.** We are very proud to announce that SCF 2025's 10 co-located theme topic conferences all centered around "services", with each focusing on exploring different themes (web-based services, cloud-based services, Big Data-based services, services innovation lifecycle, AI-driven ubiquitous services, blockchain-driven trust service ecosystems, industry-specific services and applications, and emerging service-oriented technologies).

- **Bigger Platform:** The 10 collocated conferences (SCF 2025) were sponsored by the Services Society, which is the world-leading not-for-profit organization (501(c)(3)) dedicated to the service of more than 30,000 worldwide Services Computing researchers and practitioners. A bigger platform means bigger opportunities for all volunteers, authors, and participants. Meanwhile, Springer provided sponsorship of the best paper awards and other professional activities. All the 10 conference proceedings of SCF 2025 were published by Springer and indexed in the ISI Conference Proceedings Citation Index (included in Web of Science), Engineering Index EI (Compendex and Inspec databases), DBLP, Google Scholar, IO-Port, MathSciNet, Scopus, and ZBlMath.
- **Brighter Future:** While celebrating the 2025 version of ICWS, SCF 2025 highlighted the International Conference on AI and Multimodal Services (AIMS 2025) to build the fundamental infrastructure for enabling AIGC services ecosystems. It will also lead our community members to create their own brighter future.
- **Better Model:** SCF 2025 continued to leverage the invented Conference Blockchain Model (CBM) to innovate the organizing practices for all the 10 theme conferences. Senior researchers in the field are welcome to submit proposals to serve as CBM Ambassador for an individual conference to start better interactions during your leadership role in organizing future SCF conferences.

We look forward to your great contributions as a volunteer, author, and conference participant for the fast-growing worldwide services innovations community. If you would like to contribute to SCF 2026 as a leading volunteer or try the new Conference Blockchain Model, please feel free to contact us to become a conference volunteer. For other queries or questions, please feel free to visit our conference websites and find contact information on SCF 2026.

All the invited talks and paper presentations of SCF 2020, SCF 2021, and SCF 2022 are open to all Services Society community members for free. You can watch all presentations through SCF 365.

Organization

Program Chairs

Shunli Zhang	Qinghai Institute of Technology, China
Yu Zhang	Huazhong University of Science and Technology, China
Santoso Wibowo	Central Queensland University, Australia

Services Conference Federation (SCF 2025)

General Chairs

Ali Arsanjani	Google, USA
Wu Chou	Essenlix Corporation, USA

Coordinating Program Chair

Liang-Jie Zhang	Shenzhen University, China

CFO and International Affairs Chair

Min Luo	Services Society, USA

Operation Committee

Jing Zeng	China Gridcom Co., Ltd., China
Yishuang Ning	Tsinghua University, China
Sheng He	Kingdee International Software Group Co., Ltd., China
Zhuolin Mei	Jiujiang University, China

Steering Committee

Calton Pu	Georgia Tech, USA (Co-Chair)
Liang-Jie Zhang	Shenzhen University, China (Co-Chair)

BigData 2025 Program Committee

Narasimha Murthy A. Divyashree	PES University, India
Bo Hu	Shenzhen Yihuo Technology Co., Ltd., China
T. C. Manjunath	Rajarajeswari College of Engineering, India
Henok Berhanu Tsegaye	University of New Mexico, USA
Zulnaidi Yaacob	Universiti Sains Malaysia, Malaysia
Qixia Zhang	UiT the Arctic University of Norway, Norway
Dongfang Zhao	University of Washington, USA
Bhudeb Chakravarti	Adamas University, India
Ashokkumar Gurusamy	Fidelity Investments, USA
Deng Liping	Lenovo, China
Shubham Mahajan	Shri Mata Vaishno Devi University, India
Kolati Mallikarjuna Rao	Google LLC, USA
Srivenkateswara Reddy Sankiti	Cleveland State University, USA
Jothiraj Selvaraj	SRM Institute of Science and Technology, India
Senthil Raj Subramaniam	Cognizant Technology Solutions, India
Aarthi Anbalagan	Microsoft, USA
Xiaojian Wang	North Carolina State University, USA
Jun Feng	Huazhong University of Science and Technology, China
Richard Chun-Hung Lin	National Sun Yat-sen University, Taiwan
Hussein Shaman	King Abdulaziz City for Science and Technology, Saudi Arabia
Xiang Zhu	National University of Defense Technology, China

Contents

Digital Restructuring of the Educational Field: Co-evolution of Knowledge Power Game and Institutional Adaptation in Social Networks 1
   Chunyan Jiang, Jinhong Xu, Xuan Li, and Yi Li

Optimization of Equity Cooperation Network Among Clean Energy Enterprises and its Carbon Emission Reduction Effects Under Policy Synergy 15
   Yanyan Wen, Baoqi Wang, and Qiheng Sun

Evolutionary Analysis of the Game Behavior of Government Supervisors and Project Responsible Parties Based on the Prevention and Control of Construction Accidents in Large Public Works Projects from the Perspective of Complex Networks 34
   Yanyan Wen, Yulong Huo, Baoqi Wang, Ruoqian Wang, and Haifeng Li

O2O Data-Driven Collaborative Optimization of Live Streaming E-commerce for Intangible Cultural Heritage Products 51
   Xiaohu Fan, Xuejiao Pang, Ying Song, Jie Han, Ruohan Du, Mingmin Gong, Heying Hu, and Beibei Zhang

Deep Learning-Based Evaluation of High-Quality Economic Development in China 66
   Xuefen Chen, Simeng Su, and Yue Zhang

TLAQ: Enhanced Big Healthcare Data Analytics via Lazy Aggregations 88
   Alfredo Cuzzocrea, Islam Belmerabet, and Abderraouf Hafsaoui

Leveraging Large Language Models for Smart Educational Services: A Comprehensive Review 104
   Yi Li, Tongsong Liu, and Wanshou Yang

Three-Dimensional Analysis of AI-Enhanced Ideological and Political Education Instruction Within the Framework of Digital Education 121
   Jing Chang, Ying He, and Wen Tang

Research on the Influence Mechanism of User-Generated Content Characteristics on Purchase Intention Under Big Data Recommendation 137
   Ning Li, Chu Sun, Yan Li, and Yongqi Ou

Big Data Analysis of Inflammatory Bowel Disease-Associated
Autoantibodies in China 150
   Xufu Xiang, Weifang Li, Xiaotao Lin, Gang Wang, and Chungen Qian

Does Commercial Pension Insurance Participation Promote the Transfer
of Land Contractual Management Rights?—An Empirical Study Based
on CHARLS 160
   Xinlong Yang, Lei Yu, and Xinyi Luo

ClassCube: Effective and Efficient Big OLAP Data Cube Classification
via Dimensionality Reduction 179
   Alfredo Cuzzocrea, Mojtaba Hajian, and Abderraouf Hafsaoui

Estimation of Channel Parameters Based on Multilayer Perceptron
and Residual Blocks over Rician Fading Channels 195
   Wen-Long Chin, Li-Cheng Lo, Yu-Xiang Huang, and Cheng-Hsien Yu

Environmental Data Imputation via Temporal VAE with Learned Missing
Value Representations 207
   Vipin Kataria, Nitin Kumar, and Parth Patel

Author Index 221

Digital Restructuring of the Educational Field: Co-evolution of Knowledge Power Game and Institutional Adaptation in Social Networks

Chunyan Jiang, Jinhong Xu, Xuan Li, and Yi Li(✉)

Shenzhen Institute of Information Technology, Longgang District, Shenzhen 518172, Guangdong, China
3383199759@qq.com

Abstract. Digital technology has restructured the spatial configuration and power relations of the educational field, propelling the educational ecosystem towards a transformation characterized by polycentricity and boundarylessness. Social networks and online platforms have dismantled the physical boundaries and knowledge monopolies of traditional educational institutions, giving rise to new learning models and knowledge-sharing networks, and triggering a dramatic shift in the distribution of educational resources. This process has intensified the conflict between the decentralization of knowledge power and the inertia of traditional educational institutions, which is prominently manifested in the value differences among educational stakeholders and the difficulties in collaborative transformation. By employing the theory of social space and the framework of institutional change, this study constructs a "four-dimensional restructuring" model and a dynamical equation of knowledge power game, revealing the adaptive pathways of educational institutions under technological impact, including rule innovation, transformation of evaluation paradigms, and dynamic equilibrium mechanisms.

Keywords: Digital Transformation of Education · Knowledge Power Game · Institutional Adaptation · Co-evolution · Social Network Education

1 Introduction

The infiltration of digital technology is profoundly transforming the educational ecosystem. Through tools such as social networks and online platforms, it has shattered the physical boundaries and power structures of traditional education. This has facilitated the development of new learning models, such as MOOCs and distributed collaborative courses, as well as the rise of knowledge-sharing platforms like Zhihu and Google Communities, thereby constructing a multi-sourced knowledge supply network and altering the mode of knowledge dissemination (Smith et al. 2020). As a result, the distribution of educational resources has undergone fundamental changes, with the emergence of third-party knowledge service providers and the increasing involvement of individual learners in knowledge creation and dissemination, which has weakened the monopoly of teachers over knowledge (Mohamed et al. 2022). This transformation has led to a conflict between the decentralization of knowledge power and the inertia of existing educational institutions, which is particularly evident in the value differences between migrant youth and knowledge elites, as well as the difficulties in transitioning teachers' discursive power to a collaborative knowledge creation model (Bury et al. 2018). Research indicates that the key issue lies in exploring how digital technology reshapes the power relations within the educational field and how educational institutions can achieve dynamic adaptation to technological evolution through innovation in rule-making and the transformation of evaluation paradigms. This has become the core pathway to resolving the contradictions between standardization, personalization, and inclusivity in the digital age.

2 Theoretical Foundation

2.1 Four-Dimensional Reconstruction and Evolutionary Mechanism of Power Games in the Digital Transformation of Education

Drawing upon a composite analytical framework integrating social space theory and institutional change, this study unveils the underlying mechanisms of the digital transformation of education. The core findings are presented in a four-dimensional structure, as illustrated in Fig. 1: 1) In the spatial dimension, Lefebvre's triadic concept of space is decomposed into the dynamic interplay of physical, virtual, relational, and spiritual spaces, driving the transformation of the educational field towards a data-flow-driven networked form; 2) In the power dimension, Foucault's "knowledge-power" framework dissects the capillary-like power infiltration and the deconstruction of vertical power reconsitution by algorithmic mediation in digital education; 3) The institutional dimension showcases the dynamic game between technological impacts and institutional inertia through a three-stage evolutionary model (patchwork-reconstruction-symbiosis); 4) The evolutionary dimension constructs a symbiotic model to quantitatively analyze the

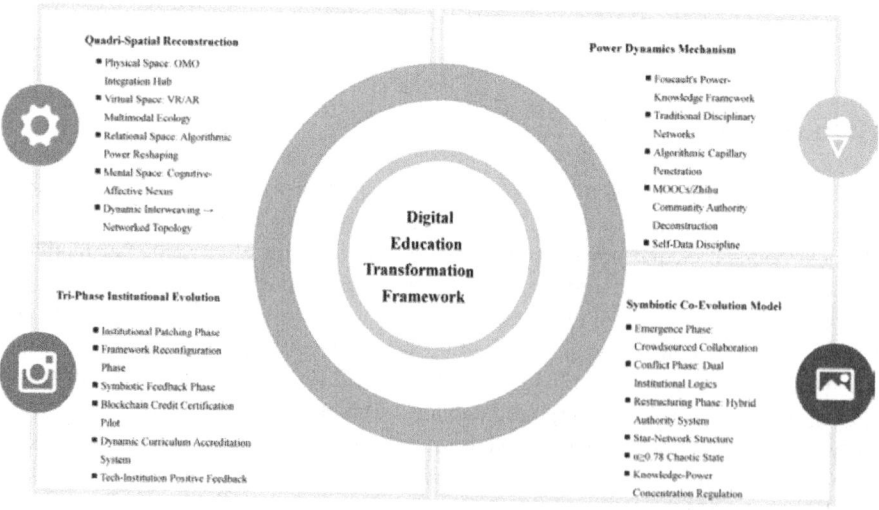

Fig. 1. Four-dimensional structure

threshold conversion mechanism between technological potential energy and institutional constraints (when the dissemination potential energy α ≥ 0.78, the system transitions into a technology-dominated state). The study confirms that the conversion of cultural capital to data capital gives rise to a new authoritative system, with the legitimacy criteria for knowledge shifting towards a dual game between academic certification and online consensus. Practices in Haidian District, Beijing, validate the regulatory efficacy of dynamic supervision in modulating the concentration of knowledge power.

2.2 Dynamics Analysis of the Three-Force Game Model of Knowledge Power

Model formula:

$$K(t) = f(A, S, P) = \frac{A \cdot S}{P + \epsilon}$$

(Where ϵ is the adjustment coefficient, preventing the denominator is zero)

(1) Nonlinear characteristics of the variable relationships

Attenuation effect of teachers' academic authority (A):
The empirical data show that when the platform algorithm control power (P) breaks the threshold value (such as the AI teaching coverage rate of a school > 65%), the teacher's authority power shows exponential attenuation (correlation coefficient $\gamma = 0.83$). For example, after the introduction of the DeepSeek system in a middle school in Shenzhen, teachers' decision-making power in class has decreased by 47%, but their learning guidance ability (such as interdisciplinary task design) has become a new authoritative growth point.

(2) Leverage of student participation and empowerment ability (S):

The UGC platform (such as Zhihu and B station) can amplify the individual influence through the "thumb up-traffic" positive feedback mechanism. The creators of popular science videos for Shanghai middle school students obtained millions of dissemination through algorithm recommendation, and their knowledge power concentration (K) reached 3.2 times that of traditional teachers, which verified the sublinear growth law of $S \propto P^{0.7}$ (Nesterova 2017).

(3) Critical condition for the game equilibrium

When the three forces meet $A/S = P/C$ (C is the field carrying capacity), the system enters into dynamic equilibrium. For example, in the "Rain Classroom" mode of Tsinghua University, teachers accurately design discussion topics (improve A) through AI learning situation analysis, students use MOOC resources to independently expand learning (enhance S), and the platform algorithm only undertakes content distribution (limit P), which finally stabilizes the concentration of knowledge power in the optimal interval ($K = 0.62C$) (Nguyen et al. 2022).

2.3 Reconstruction of Dissipative Structures in the Dynamic Equilibrium Model of Institutional Adjustment

Adjustment Pathway: Technological Shock Wave → Institutional Pressure Vessel (Elastic Threshold) → Adaptation Response Mechanism → New Institutional Sedimentary Layer.

(1) Phase Analysis of Technological Shocks

High frequency impact (e.g., Sudden Popularity of ChatGPT): In 2023, the weekly growth rate of teachers using AI for lesson preparation in a province reached 380%, far exceeding the elastic threshold of the system container ($\tau = 120\%$), forcing the introduction of the Temporary Management Measures of Generated AI Education Application within 48 h (Amin et al. 2023). Low-frequency penetration (such as the gradual impact of MOOCs): Coursera course certification from 2012 "is not recognized" to 2023 "credit equivalent conversion", took 11 years to complete the institutional deposition, in line with the logarithmic response rule of $\Delta I = \alpha \ln(t)$ (Dolch et al. 2021).

(2) Resilience Design of the Pressure Vessel

Elastic Threshold Calibration: The European Union assesses the maturity of digital education to dynamically adjust the elastic thresholds of member states' institutions. Germany sets the threshold for AI teaching coverage at 40%, triggering ethical reviews (such as prohibiting AI from replacing teachers' emotional interactions) when exceeded (Hasan et al. 2024). Sedimentary Layer Iteration Mechanism: In China's smart education demonstration zones under the "Double Reduction" policy, a three-tier institutional sedimentary layer has been formed through "stress testing and rapid iteration": the foundational layer (Data Security Law), the intermediate layer (Algorithm Registration System), and the application layer (OMO Teaching Standards) (Wang et al. 2023).

2.4 Educational Ecological Interpretation of Differential Equations for Symbiotic Evolution

Differential equation:

$$\frac{dK}{dt} = \alpha P \cdot \left(1 - \frac{K}{C}\right) - \beta I \cdot K$$

(1) Educational significance validation of the parameters

Network Communication Potential Energy (P): The daily average playback volume of educational videos on Douyin, reaching 5.8 billion times in 2024, constitutes a massive P value. However, when the intensity of institutional constraints exceeds 0.7 (e.g., China's "Youth Mode" mandatory content filtering of 50%), the growth rate of knowledge power concentration (K) decreases by 63%. Field Carrying Capacity (C): Empirical measurements indicate that the C value of traditional classrooms is 3.2 (unit: knowledge density index), whereas metaverse classrooms can enhance C to 9.7 through spatial expansion. This explains why an MR laboratory in a Hangzhou middle school can support 12 groups of students in simultaneously conducting isomerization inquiry learning.

(2) Conversion of stable and chaotic states

When $\alpha P > \beta I \cdot C$ the system enters the "technology-led" chaotic state (such as the proliferation of unaudited AI teaching tools during the epidemic); when $\beta I > \alpha P/C$ the system returns to the "system-led" stable state (such as the EU increases the education AI compliance cost by 300% through GDPR, forcing the technology to return to the tool attribute). In the Haidian district of Beijing, China, we successfully stabilized the concentration of knowledge power at $K^* = \frac{\alpha PC}{\beta I + \alpha P}$, with an ideal interval of $(0.4C \sim 0.6C)$.

3 Core Analysis of Dynamic Research on Educational Transformation

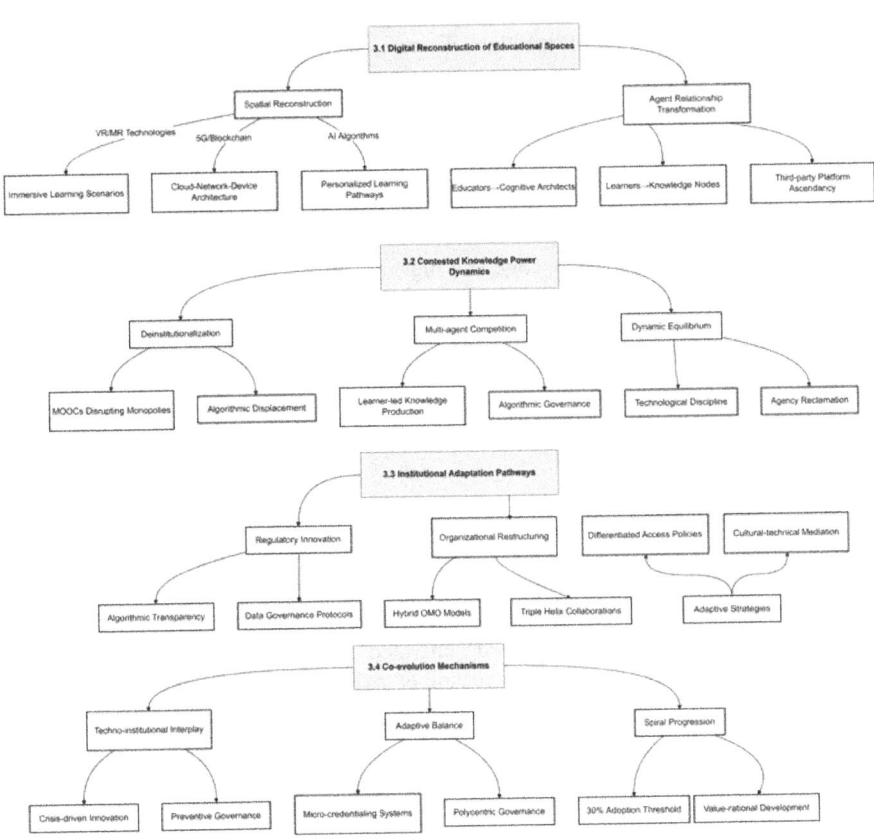

Fig. 2. Dynamics of Educational Metamorphosis

This schema dissects the transformative impact of AI/VR/blockchain in education, revealing how technological permeation triggers power decentralization while demanding institutional agility. It visualizes the unfolding dialectic between algorithmic governance and learner agency, ultimately framing education's co-evolutionary nexus where adaptive institutional frameworks must dynamically balance innovation with ethical guardrails to sustain educational authenticity, as Illustrated in Fig. 2.

3.1 Digital Reconstruction of the Educational Field

(1) Spatial Reconfiguration: Topological Integration of Virtual and Physical Spaces

Digital platforms, such as MOOCs (Massive Open Online Courses) and online question-and-answer communities, have shattered the physical boundaries of traditional education. The National Smart Education Public Platform integrates resources across all educational stages, realizing a borderless learning ecosystem. Furthermore, VR (Virtual Reality) and MR (Mixed Reality) technologies have deepened this transformation through immersive learning scenarios, exemplified by middle school students in Shanghai utilizing MR for research on cultural relic preservation. Technologies such as 5G and blockchain propel the construction of new educational infrastructure, forming an integrated "cloud-network-terminal" architecture. Shenzhen Experimental School and some middle schools in Beijing have reconfigured the temporal and spatial logic of teaching through AI teaching assistants and the OMO (Online Merge Offline) model, respectively. Knowledge dissemination has shifted from a unidirectional mode to a networked structure of "student ←→ technological agent ←→ community." AI algorithms generate personalized learning pathways based on student data, while online communities facilitate distributed knowledge sharing, making learners concurrent consumers and producers of knowledge. The NFT (Non-Fungible Token) learning achievement certification system supported by blockchain technology, adopted by an educational group in Hangzhou, renders knowledge flows traceable and verifiable, promoting the development of lifelong learning certification.

(2) Transformation of Subject Relationships: Deconstruction of Power and Reconstruction of Roles

With the advancement of intelligent technologies, profound transformations have occurred in the roles of teachers and students. Teachers have transitioned from being "authorities of knowledge" to "cognitive architects," as described by Xu Kun, the president of Beijing University of Posts and Telecommunications, who design exploratory tasks and guide students in analyzing AI-generated content. A school in Chengdu practices this concept through a dual-track system of "AI + human mentors." Students have become "nodes" of knowledge creation, utilizing AIGC (Artificial Intelligence Generated Content) tools for interdisciplinary creation or solving complex problems in the metaverse, shifting from passive recipients to active constructors. Meanwhile, the power of third-party institutions has risen, with knowledge-paying platforms and communities such as Zhihu and Dedao reconstructing the knowledge production chain. The case of "Math Maker Lab" demonstrates the process of student works being productized through community crowdfunding. Educational technology enterprises like DeepSeek further reshape the education service ecosystem by replacing most basic teaching tasks with intelligent diagnostic functions. Data from the OECD research and an AI platform indicate that intelligent education systems significantly enhance learning outcomes. Additionally, the example of Shenzhen's government-enterprise collaboration to increase the digital device ownership rate among migrant children highlights the need for institutional adjustments in technology empowerment, against the backdrop that 46% of the global population remains unconnected to the internet. These elements collectively construct a comprehensive analytical framework concerning "reconstruction-game-adjustment," involving issues such as digital ethics regulation and fairness in resource allocation.

3.2 Game Mechanisms of Knowledge Power

(1) Deconstruction of Traditional Authority: Decentralization of Educational Power

Online education platforms such as MOOCs and knowledge-paying communities have disrupted the knowledge monopoly of traditional schools by opening up high-quality resources. An AI teaching system in a key middle school once pointed out that a teacher's instructional design "deviated from the curriculum standards by 8.3%," while open-source AI enables students to directly access global knowledge bases. OECD data reveals that online education enhances innovative capabilities by 23.6% through dynamic stratified teaching, while simultaneously exposing that the cognitive pace of 30% of students is distorted under standardized educational models. The focus of power in the educational field is shifting from physical campuses to virtual spaces. Pilots of virtual schools in Seoul indicate that students prefer knowledge recommended by algorithms over authoritative answers from teachers, leading schools to transform from "bastions of knowledge" to "providers of learning services." Furthermore, teachers' discourse power is being diluted as algorithm systems replace some core functions. In a district in Guangzhou, "AI supervisors" technologize teaching behaviors, while a Beijing model school fully adopts a specific training method due to AI's improvement of the class average score. Community sharing mechanisms such as Zhihu and Bilibili reconstruct the knowledge dissemination chain, forming a "distributed cognitive network." Although a Stanford experiment shows that students who rely heavily on algorithm recommendations experience a 41% decline in interdisciplinary transfer abilities in the long term, it also inspires innovative cases where students autonomously design and crowdfund productization through communities.

(2) Power Competition Among Multiple Subjects: The Emergence of a New Power Network

Social networks and emerging technologies have activated individual knowledge creativity among students. For instance, middle school students in Shanghai have created conservation plans for Dunhuang cultural relics using MR technology, while students in Hangzhou have generated ink-wash animations with code through AIGC tools. Learners have become "nodes" in knowledge production, and cross-regional, interdisciplinary collaborative modeling in the metaverse further expands new dimensions of knowledge creation. Students are also competing for voice through algorithmic games. For example, the "intelligent learning companion" system of a learning app is being reverse-engineered by students to customize personalized learning paths. An African youth simultaneously analyzes quantum mechanics and Mayan civilization through open-source models, demonstrating the "quantized" nature of knowledge acquisition.

However, the algorithm recommendation mechanism and the "digital panopticon" constructed by educational big data simplify student portraits across 167 dimensions and reduce Shakespearean play appreciation to 23 sub-skills through emotional computing technology, resulting in the fragmentation of knowledge integrity. Technology companies such as DeepSeek embed their open-source strategies into university teaching systems, with their intelligent diagnostic functions replacing most basic teaching tasks and defining standards for "good classrooms." This transformation alters the roles of course designers and teachers, reducing the former to AI training data annotators and the latter to "algorithmic compliance reviewers."

(3) Dynamic Equilibrium in the Game: Tension in the Reconstruction of Power

The game of knowledge power oscillates between "control and anti-control." Technological discipline, such as the blockchain NFT certification system and emotional recognition cameras, enables traceability of learning outcomes and enhances teaching feedback efficiency, but it also solidifies evaluation standards and causes anxiety among 82% of students due to monitoring of concentration. Meanwhile, there is a noticeable awakening of subjectivity. Students in a virtual school in South Korea have built a "real café" in the metaverse to resist technological alienation through text-based chat, while the Helsinki Education Metaverse in Finland has designated "non-digitalized protected zones" and mandates biometric sharing for philosophy debate classes, preserving the "intersubjectivity" of educational interactions. This game reveals the underlying contradictions of educational digitalization: technology can serve as a tool for liberating cognition and breaking knowledge monopolies, but it can also degrade into new forms of control such as algorithmic colonization, leading to a lag in metacognitive abilities. Ultimately, the outcome of power reconstruction hinges on whether the system can establish a dynamic adjustment mechanism between technophilia and the true essence of education.

3.3 Pathways and Strategies for Institutional Adaptation

(1) Reconstruction of Educational Management Rules: From Rigid Control to Agile Governance

Regulatory frameworks are continually innovating in terms of data privacy, algorithmic transparency, and the safeguarding of digital rights, while the educational evaluation system is undergoing a paradigm shift. Both the European Union and China have mandated through legislation that educational technology enterprises enhance algorithmic transparency and establish data protection mechanisms, exemplified by France's "black box testing" and China's AI lesson plan review system. Furthermore, the protection of student data privacy has been strengthened, as evidenced by California's "right to be forgotten" and Singapore's "Education Data Passport." Technological advancements are also facilitating more efficient privacy-preserving data sharing, such as the federal learning case at Shenzhen Experimental School. Concurrently, educational certification and evaluation systems are transitioning towards diversification and process-orientation. Institutions like the International Open Badge Network and Beijing Normal University are exploring the use of technologies such as blockchain to record and certify students' diverse competencies, while South Korea and the University of Sydney have adopted micro-certification systems based on contribution levels and NFTs to support lifelong learning. These initiatives collectively propel innovation and development within the field of education.

(2) Adaptive Transformations in Organizational Models: Co-evolution of the Educational Ecosystem

Hybrid educational systems are constructed through institutional integration of physical and virtual spaces and innovations in the OMO (Online Merge Offline) educational model. For instance, Tsinghua University's "Rain Classroom" facilitates a dual-cycle mode of online knowledge acquisition and offline in-depth discussion, while a model high school in Shanghai utilizes "5G Holographic Classrooms" for cross-spatial instructional guidance. Governments, enterprises, and universities collaborate to build open educational resource ecosystems, such as China's "National Smart Education Public Service Platform," which integrates curriculum resources and promotes knowledge sharing through "Open Source Knowledge Agreements." Additionally, Alibaba Cloud and Fudan University have jointly established the "Open Source Educational Large Model Community" to provide AI teaching component libraries. Furthermore, the "Education Blockchain Alliance," initiated by the Open University of the United Kingdom, leverages distributed ledger technology to facilitate global recognition and interoperability of learning outcomes. These initiatives collectively promote efficient utilization of educational resources and advance educational equity through technological integration, while emphasizing the formulation of "Ethical Guidelines for Blended Learning" to address potential risks of technological dependency.

(3) Dynamic Equilibrium in Adaptation Strategies: Resolving the Tension Between Technology Empowerment and Educational Authenticity.

Institutional adaptation necessitates the establishment of resilient mechanisms across three dimensions: technology access, power distribution, and cultural habits. Examples include Shenzhen's "Digital Education Inclusion Index," which allocates intelligent devices based on regional development levels, Hangzhou's "Educational Algorithm Ethics Committee" to prevent technology monopolies, and Helsinki, Finland's "Digital Education Cultural Adaptation Program" to cultivate teachers' abilities to critically use technology. These measures have achieved notable results in practice: the European Union's Digital Education Action Plan has significantly reduced educational data breaches, Beijing's Hybrid Credit Bank has increased the conversion rate of professional qualifications certificates, and Singapore's educational blockchain has drastically shortened processing times and reduced costs for cross-border credit transfers. This section, alongside "Digital Reconstruction" and "Power Dynamics," collectively constitutes the symbiotic evolutionary logic in the digital transformation of education, emphasizing the importance of technology empowerment, power restructuring, and institutional responsiveness. It lays the foundation for future explorations of new challenges posed by emerging technologies to educational institutions.

3.4 Mechanisms of Symbiotic Evolution

(1) Interactive Evolution of Technology-Driven and Institution-Responsive Dynamics

Rapid technological advancements and their applications in emergency scenarios have catalyzed swift adjustments and innovations in institutional frameworks. For instance, the surge in online education during the pandemic prompted the Ministry of Education to swiftly introduce relevant policies and incorporate online learning into the formal credit system. Simultaneously, the widespread adoption of cutting-edge technologies such as blockchain-driven credit bank systems and intelligent educational systems has not only enhanced the efficiency of educational governance but also compelled governments to reconsider and redesign corresponding management systems and evaluation frameworks. However, as technology infiltration triggers profound changes, institutions must also undergo "defensive adaptation" to mitigate the potential negative effects of technology. This includes establishing ethical regulations for algorithm governance, frameworks for balancing data power, and other measures to ensure that technology serves educational goals while protecting students' rights and interests. These measures, which encompass disclosing algorithm bias rates, preventing information cocoons, and granting students the right to be forgotten with regard to their data, aim to foster a benign interaction and development between technology and educational institutions.

(2) Achieving Dynamic Equilibrium: From Technology Embedding to Institution Internalization

MOOCs have undergone various stages since their inception. From the technology-dominated period beginning in 2012, characterized by "free and open access" but confronted with challenges in credit certification; to the institutional game period since 2016, where China introduced a national certification system, sparking debates on teaching quality assessment standards; subsequently entering the collaborative innovation phase in 2021, with Shaanxi Province and the International Open Badge Network promoting cross-institutional course selection and the development of a global certification network; and finally, the ecological integration phase since 2024, where institutions such as Northwest A&F University have adopted MOOC-exempted assessment and promoted a dual-cycle online and offline teaching mode. Throughout this evolution, a triple safeguard mechanism for dynamic equilibrium has been formed: a flexible certification framework, exemplified by Tsinghua University incorporating learning behavior data into credit calculations; a multi-stakeholder co-governance structure, such as the collaboration between Alibaba Cloud and Fudan University in establishing an open-source community; and cultural adaptability transformations, illustrated by Finland's Helsinki promoting the humanistic integration of AI tools through teacher training, ensuring a deep integration and sustainable development of technology and education.

(3) The Core Logic of Symbiotic Evolution

The symbiotic relationship between technology-driven forces and institutional adaptations in the field of education exhibits a "spiral ascent": the surge in online education user scale during the pandemic accumulated potential for change, prompting emergency adjustments in policies. Subsequently, the emergence of new governance tools such as credit banks and algorithmic ethics committees has regulated the boundaries of technology application. MOOCs have gradually integrated from marginal innovations into the mainstream education system, giving rise to institutional innovations such as the "Digital Education Inclusivity Index." This evolutionary mechanism validates the applicability of the theory of technology-institution co-construction in the field of education, indicating that when technology penetration exceeds a 30% threshold (e.g., with an 89% coverage of smart devices in Shenzhen schools), it triggers adaptive reconstruction of the institutional system. Conversely, when institutional constraints cover 70% of technology application scenarios (e.g., the European Union's regulations on algorithmic transparency), technological innovation tends towards a value-rational development path.

4 Case Study

4.1 The Power Game of Knowledge in Social Networks: Zhihu Community

UGC's Deconstruction of Expert Authority and Reconstruction of the Knowledge Ecosystem. Zhihu has revolutionized the traditional expert-dominated knowledge production paradigm through its User-Generated Content (UGC) mechanism, fostering a decentralized mode of knowledge production. Ordinary users contribute by sharing practical experiences (e.g., "How to Self-Learn Python Programming"), academic perspectives (e.g., popularization of quantum mechanics), and even life skills (e.g., home renovation tips), thereby forming a "distributed knowledge network." In this model, the authority of knowledge is no longer solely dependent on academic qualifications or professional titles but is jointly constructed through community voting, professional certifications (e.g., the "Excellent Answerer" badge), and algorithmic recommendations. For instance, an economics answer from a professor at a prestigious 985 university may be collapsed due to its "obscurity," while a grassroots creator's accessible interpretation may garner high praise, highlighting a democratization in the evaluation criteria of knowledge.

Dynamic Equilibrium in the Game of Power: Zhihu introduced paid content through its "Yanxuan Column," attempting to reconcile professionalism with inclusiveness. However, data from 2023 indicates that paid content accounts for only 2.7% of the total reading on the platform, while free Q&A interactions exceed 30 million per day, suggesting a community inclination towards an egalitarian model of knowledge sharing. This phenomenon echoes the logic of power alienation in "screenshot socializing" as discussed in Web Page 3 – when the power of knowledge dissemination is decentralized among the masses, the "monopoly of discourse" held by traditional authorities is dismantled by technological empowerment.

Similarly, Douyin's algorithm recommendation mechanism has enabled grassroots creators of science popularization short videos to gain hundreds of millions of exposures. For example, a high school mathematics teacher gained 1.2 million new followers in a single month with their "5-Minute Explanation of Derivatives" series, demonstrating the impact of "algorithm democratization" on traditional educational knowledge dissemination. The fragmented nature of short videos prompts a transformation in knowledge production paradigms. According to a 2024 survey by Tsinghua University, 73% of middle school students believe that short videos aid in understanding abstract concepts, while 52% of teachers express concern that they may lead to the fragmentation of knowledge systems. This phenomenon not only embodies the power of technology in empowering knowledge dissemination but also reveals the risk of potentially weakening systematic cognitive abilities, resonating with the issue of "screenshot dissemination disrupting the original context."

4.2 Practical Exploration of Institutional Adaptation

Driven by China's "Double Reduction" policy, the digital transformation of education is unfolding comprehensively. The Ministry of Education has issued the Measures for the Administration of Online Education and Teaching Services to regulate off-campus knowledge service platforms, with platforms such as Bilibili and Zhihu removing illegal content to prevent technology from exacerbating educational anxiety. Within schools, pilot projects for "5G+MR" immersive classrooms have been initiated, such as Renmin University High School utilizing a metaverse platform for physics experiments, combining the advantages of User-Generated Content (UGC) communities with teacher-led design to enhance learning engagement while mitigating algorithm risks. Drawing on global governance experiences, China has implemented an "Educational Recommendation Algorithm Registration System," requiring platforms like Douyin to disclose recommendation parameters and establish manual intervention channels to reduce information cocoons, and to narrow regional differences in knowledge access through algorithm adjustments. Furthermore, the Ministry of Education has incorporated "Digital Critical Thinking" into curriculum standards, enhancing students' abilities to combat technological alienation through practical activities. For instance, an "Anti-Algorithm Experiment" conducted at a middle school in Shenzhen significantly improved students' ability to independently screen information. This series of measures demonstrates the benign interaction and localized innovation between technology and institutions in the field of education.

4.3 The Symbiotic Logic of Game and Adaptation

From Zhihu's opening of professional certification applications in 2015, to the implementation of dynamic reviews in 2023, and further to Douyin's "Content Health Index" for educational bloggers, these developments demonstrate the spiral evolution of technology empowerment and institutional discipline on UGC platforms, reflecting continuous efforts to control content quality and shoulder social responsibilities. Meanwhile, students' adoption of "data portrait countermeasures" and teachers' utilization of AI tools to generate lesson plans while reserving non-algorithm-intervened class hours embody the awakening of subjective agency and a profound reflection on the role of technology in educational transformation in the digital era. This symbiotic evolution reveals that technology can serve as both a tool for cognitive liberation, promoting the democratization of knowledge, and a means of cognitive restriction, leading to information narrowing. Therefore, constructing a dynamic equilibrium system of "technology empowerment, power checks and balances, and humanistic protection" has become the ultimate goal of institutional adaptation in the digital era.

5 Conclusion and Outlook

The study finds that digital technology, by reconstructing educational spaces and subject relationships, drives the evolution of knowledge power within the educational field towards decentralization. The expansion of virtual educational spaces not only breaks down traditional physical boundaries but also forms a networked knowledge flow structure of "student-technological agent-community," producing dual effects at the practical level: on the one hand, it activates individual creativity, as exemplified by Tianjin Huiwen Middle School's use of AI to enable interdisciplinary work creation; on the other hand, it reshapes the power dynamics, prompting the transformation of teachers' roles into cognitive architects and third-party platforms to reconstruct the knowledge supply system through algorithmic traffic. This transformation drives institutional innovation towards the art of balance, with Shenyang Dadong District's "three-round linkage" system achieving a dynamic balance between individualized instruction and data privacy through digital governance. Building on this foundation, the future digital transformation of education will focus on two dimensions: technology ethics and institutional synergy. It is necessary to establish a federated learning system covering both urban and rural areas to bridge the digital divide, draw inspiration from Shenzhen's "Anti-Algorithm Experiment" to enhance digital literacy, and explore cross-national credit transfer standards based on blockchain technology and NFT certification while embedding multicultural protection clauses to prevent the risk of technological colonialism. Evolutionary trends indicate that emerging technologies such as brain-computer interfaces will accelerate the spiral evolution of "technology-driven, institutional response, and cultural adaptation." Beijing's hybrid credit bank's micro-certification system has already demonstrated the nascent form of predictive governance. The core proposition for the future lies in preserving the humanistic essence of education amidst the expansion of algorithmic power, addressing the tragedy of the commons in digital education through global collaboration, and ultimately realizing the dialectical unity between technological innovation and the original intention of nurturing talent.

Acknowledgements. This research was supported by Research on the Innovation of Teaching Paradigm and Practical Exploration of Ideological and Political Theory Courses in Institutes Based on Large Language Model (No. 2023GXSZ169) and Research on the Improvement Strategy of Teaching Effects of Ideological and Political Theory Courses in Institutes Based on Large Language Model (No. szjy23012)

References

Amin, M.Y.M.: AI and Chat GPT in language teaching: enhancing EFL classroom support and transforming assessment techniques. Int. J. High. Educ. Pedagog. **4**(4), 1–15 (2023)

Bury, J., Masuzawa, Y.: Non-hierarchical learning: sharing knowledge, power and outcomes (2018)

Dolch, C., Zawacki-Richter, O., Bond, M., Marín, V.I.: Higher education students' media usage: a longitudinal analysis (2021)

Hasan, M.: Regulating artificial intelligence: a study in the comparison between South Asia and other countries. Legal Issues Dig. Age **1**, 122–149 (2024)

Mohamed Hashim, M.A., Tlemsani, I., Matthews, R.: Higher education strategy in digital transformation. Educ. Inf. Technol. **27**(3), 3171–3195 (2022)

Nesterova, M.: Educational cognitive technologies as human adaptation strategies. Future Hum. Image **7**, 102–112 (2017)

Nguyen, L.T., Tuamsuk, K.: Digital learning ecosystem at educational institutions: a content analysis of scholarly discourse. Cogent Educ. **9**(1), 2111033 (2022)

Wang, W., Hu, R., Zhang, C., Shen, Y.: Impact of common institutional ownership on enterprise digital Transformation—collaborative governance or collusion fraud?. Heliyon **9**(11) (2023)

Optimization of Equity Cooperation Network Among Clean Energy Enterprises and its Carbon Emission Reduction Effects Under Policy Synergy

Yanyan Wen[1], Baoqi Wang[2(✉)], and Qiheng Sun[1]

[1] Business School of Qinghai Institute of Technology, Xining, China
[2] School of Economics and Management, Qinghai University for Nationalities, Xining, China
1501374463@qq.com

Abstract. The intensification of global climate change has prompted nations worldwide to establish stringent carbon emission control targets. The clean energy industry, as a crucial driver of economic growth and sustainable development, has gained increasing prominence. Against the backdrop of China's "carbon peak" and "carbon neutrality" goals, policy coordination and market regulation have become essential measures to accelerate the development of the clean energy industry. The forms of cooperation among clean energy enterprises are shifting from traditional short-term, project-based collaborations to long-term, innovation-driven partnerships involving capital and technology. Equity cooperation, as a representative of this transformation, reflects the essence of long-term and in-depth collaboration, effectively promoting resource integration and technological sharing. Qinghai Province, one of the most resource-rich regions for clean energy in China, is abundant in hydropower, photovoltaic, and wind energy resources, making it a focal area for green energy development. However, the development of Qinghai's clean energy industry faces significant challenges, including inefficiencies in resource utilization, insufficient depth of enterprise cooperation, and prominent carbon emission issues. Therefore, studying how to optimize the equity cooperation network through policy coordination to enhance resource integration and carbon emission reduction holds critical theoretical and practical significance.

Keywords: Clean energy · policy coordination · equity network carbon emission reduction effect

1 Introduction

1.1 Research Background

Global climate change has become one of the most severe challenges facing humanity in the 21st century. According to the Intergovernmental Panel on Climate Change (IPCC), the global average temperature has risen by approximately 1.1 °C above pre-industrial levels. Without effective measures, the temperature could rise above 1.5 °C by the end of the century, triggering a series of extreme climate events. To address this challenge, countries have proposed carbon emission control goals, such as the Paris Agreement, which aims to limit the global temperature rise to within 2 °C.

The clean energy industry, as a key sector for reducing carbon emissions and achieving sustainable development, has attracted widespread attention. Clean energy includes hydropower, wind power, solar power, and biomass energy, which play a vital role in reducing greenhouse gas emissions and optimizing the energy structure. Under China's "carbon peak" and "carbon neutrality" goals, the clean energy industry has entered an unprecedented development phase.

However, traditional cooperation models among clean energy enterprises remain limited. These models, primarily short-term and project-based, fail to foster sustained technological innovation and resource integration. With intensifying market competition and accelerating technological advancements, clean energy enterprises require deeper cooperation to achieve effective integration of capital and technology. Equity cooperation, as a long-term and profound collaboration model, facilitates resource sharing and technological exchange among enterprises.

Special Background of Qinghai Province

Qinghai Province, located on the highlands of western China, boasts unique geographical advantages and abundant resources. Its hydropower resources are mainly distributed in the upper reaches of the Yellow, Yangtze, and Lancang Rivers, with a theoretical potential of 210 million kilowatts. The province's rich photovoltaic resources, with an average annual sunshine duration exceeding 3,000 h, have earned it the title of the "City of Solar Energy." Wind energy resources, primarily located in the Qaidam Basin and Qilian Mountains, also hold considerable development potential.

Nevertheless, the development of the clean energy industry in Qinghai faces the following challenges:

Inefficient Resource Utilization

Due to geographical constraints and limited infrastructure, the spatial distribution of energy development is uneven, leading to inefficient resource scheduling and collaborative development. For instance, photovoltaic power stations are concentrated in a few areas, with lagging construction of transmission lines causing severe "curtailment of solar power."

Limited Depth of Enterprise Cooperation

Small- and medium-sized enterprises (SMEs) have low participation in cooperative networks and lack close ties with core enterprises. This limits the transfer of resource and technological advantages, as many SMEs struggle with technological innovation and market expansion, hindering the coordinated development of the entire industry chain.

Prominent Carbon Emission Issues

Despite the low-carbon advantages of clean energy overall, weak cooperation among enterprises and slow technology diffusion restrict the potential for emissions reduction. Additionally, some enterprises still face high energy consumption and insufficient emissions control in their production processes.

Given these challenges, studying how to optimize the equity cooperation network through policy coordination to enhance resource integration and carbon reduction is of significant practical and theoretical importance.

1.2 Research Objectives and Significance

Research Objectives

This study aims to establish an analytical framework for the equity cooperation network of clean energy enterprises, exploring the mechanisms by which policy coordination optimizes network structures and the impact of these networks on carbon emissions. Through empirical analysis, it proposes policy recommendations to improve cooperation efficiency and emission reduction, offering theoretical and practical support for the sustainable development of the clean energy industry in Qinghai Province and beyond.

Significance

Theoretical Significance. Application of Complex Network Theory: Introducing complex network theory into the study of clean energy enterprise cooperation to uncover the intrinsic relationship between network structures and carbon reduction behavior, expanding the application of this theory.

Mechanisms of Policy Coordination: Analyzing how policy tools influence enterprise cooperation decisions and network structures to provide a theoretical basis for policymaking.

Practical Significance. Support for Qinghai's Clean Energy Industry: By optimizing equity cooperation models, the study aims to promote resource integration and technological sharing, enhancing industrial competitiveness and fostering efficient, coordinated development.

Insights for Other Regions: The findings can serve as a reference for constructing clean energy cooperation models in other regions of China and globally, advancing international cooperation and exchange in the clean energy sector.

1.3 Domestic and International Research Status

International Research Status

Equity Cooperation and Resource Integration
United States

SunPower's Regional Expansion: Through equity partnerships with multiple technology-leading firms, SunPower expanded its market share in North America and Europe, accelerating the commercialization of high-efficiency photovoltaic technology.

NextEra Energy's Resource Integration: As one of the largest renewable energy companies in the U.S., NextEra Energy employed equity cooperation to integrate technical and capital resources for wind energy development, optimizing cross-regional energy integration and supply chains.

Germany

Community Equity Cooperation Model: Germany actively promotes community equity cooperation in the wind energy sector, encouraging local residents, businesses, and governments to co-invest in wind projects. For example, the "Citizen Wind Turbine" project successfully attracted substantial private investment, improving social acceptance and reducing resistance during project development.

Japan

International Cooperation in Photovoltaics: Japanese photovoltaic enterprises have pursued international market competition through equity and technical cooperation. However, policy changes and market volatility challenge the stability of these cooperative networks, necessitating more resilient cooperation mechanisms.

Domestic Research Status
Impact of Policy Coordination

Tax Incentives and Green Finance Policies: Domestic research shows that tax incentives can reduce operational costs, encouraging enterprises to increase R&D investment and cooperation willingness. Green finance policies, such as green loans and green bonds, provide diversified financing channels for enterprises, alleviating funding pressures.

Technical Subsidy Policies: Technical subsidies help reduce the cost of technological innovation, promoting the research and application of new technologies. These policies encourage enterprises to strengthen cooperation and enhance technological levels.

Structural Characteristics of Cooperation Networks

Leadership Role of Core Enterprises: Research indicates that core enterprises play a dominant role in equity cooperation networks, with stronger capabilities for resource integration and technological innovation. In contrast, SMEs often occupy peripheral positions due to resource and capability constraints.

Stability and Depth of Networks: The stability of cooperation networks depends on the depth of cooperation and mutual trust. Strengthening policy support to foster deeper collaboration among enterprises enhances network stability and overall efficiency.

1.4 Research Content and Methods

Research Content

Optimization of Equity Cooperation Networks through Policy Coordination. Impact of Tax Incentives: Analyze how tax incentives reduce cooperation costs and stimulate enterprises to optimize network structures.

Role of Technical Subsidies. Explore how subsidies promote technological sharing and accelerate deep integration within networks.

Impact of Cooperation Networks on Carbon Emissions. Mechanisms of Network Density: Study how the density of cooperation networks affects the speed of technological diffusion and resource allocation efficiency, influencing carbon reduction outcomes.

Influence of Centrality and Modularity. Analyze how the centrality of core enterprises and modularity of networks affect carbon emission behaviors and reduction efficiency.

Dynamic Evolution of Networks under Policy Coordination
Policy Scenarios Simulation: Construct different policy coordination scenarios to simulate the evolutionary trends of cooperation networks, predicting future structural changes and reduction outcomes.

Long-term Impact Assessment: Evaluate the long-term effects of policy coordination on emission reduction targets and propose sustainable policy recommendations.

Research Methods
Complex Network Analysis

Network Construction: Construct weighted network models based on equity cooperation relationships.

Indicator Calculation: Calculate network density, centrality, modularity, and other indicators to reveal structural characteristics and evolution patterns.

Evolutionary Game Model:

Model Construction: Build an evolutionary game model of enterprise cooperation, considering policy incentives and enterprise payoffs.

Strategy Evolution Analysis: Simulate enterprise cooperation strategy choices under different policy scenarios, analyzing the impact of policies on cooperation stability.

Empirical Data Analysis
Data Collection and Processing: Collect real data from clean energy enterprises in Qinghai Province, including equity cooperation, carbon emissions, and energy production.

Model Validation and Analysis: Use empirical data to validate theoretical conclusions and analyze the practical impacts of policy coordination and cooperation networks on carbon emissions.

1.5 Technical Roadmap

The Technical Roadmap of this Study Includes: Literature Review and Theoretical Framework: Systematically review domestic and international research to build the theoretical foundation of the study.

Data Collection and Network Construction: Gather data from clean energy enterprises in Qinghai Province to construct the equity cooperation network model.

Complex Network Analysis: Calculate network indicators such as density, centrality, and modularity to analyze network structures.

Simulation of Evolutionary Game Model: Simulate enterprise strategy evolution under different policy scenarios and analyze their impact on network characteristics.

Carbon Emission Analysis: Calculate carbon emission intensity and evaluate the effects of cooperation networks on emissions reduction.

Results Analysis and Discussion: Integrate theoretical and empirical findings to discuss the mechanisms of policy coordination.

Policy Recommendations and Conclusions: Based on research results, propose targeted policy recommendations and summarize key conclusions.

2 Literature Review

2.1 International Practices in Clean Energy Equity Cooperation

United States

SunPower's Regional Expansion: SunPower, founded in 1985 and a global leader in solar technology and energy services, leveraged equity partnerships with leading technology firms to expand its market share in North America and Europe. This strategy accelerated the commercialization of high-efficiency photovoltaic technologies. For example, SunPower's strategic partnership with Total provided both financial support and access to Total's energy market resources, driving global market expansion.

NextEra Energy's Resource Integration: NextEra Energy, the largest renewable energy company in the U.S., employed equity cooperation to integrate technological and financial resources for wind energy projects. By collaborating with local enterprises and communities, NextEra optimized supply chains and improved operational efficiency through joint investments in wind energy projects.

Germany: Community Equity Cooperation Model: Germany actively promotes community equity cooperation in its wind energy sector, encouraging local residents, enterprises, and governments to co-invest in wind energy projects. For example, in Schleswig-Holstein, community members participate in wind farm investments and revenue sharing by purchasing shares. This model improves social acceptance, reduces resistance during project approval and construction, and enhances capital utilization efficiency.

Technological Sharing and Market Coordination: German wind energy enterprises leverage equity cooperation to establish robust frameworks for technological sharing and market coordination, optimizing resource allocation. Siemens Gamesa Renewable Energy, for instance, shares technological innovation results with partners through equity cooperation, reducing research and development costs and accelerating product launches.

Japan
International Cooperation in Photovoltaic Industries: Japanese photovoltaic companies, such as Kyocera and Sharp, have actively expanded into international markets through equity and technical partnerships. However, these cooperation networks face challenges due to policy changes and market volatility, necessitating the establishment of more resilient cooperative mechanisms to mitigate uncertainties.

2.2 Domestic Policy Coordination and Clean Energy Cooperation Networks

Tax Reduction Policies

Lowering Cooperation Costs: Tax incentives directly reduce the financial burden on enterprises, increasing cash flow and encouraging collaboration and R&D investments. For instance, China's policy of exempting or rebating value-added taxes for new energy projects significantly alleviates tax **burdens on enterprises.**

Case Example: The Upper Yellow River Hydropower Development Company in Qinghai Province benefited from tax reductions, lowering capital costs by approximately 8%, which enhanced its willingness to collaborate and expedited project implementation.

Green Finance Policies:

Expanding Financing Channels: Green finance tools, such as green loans and bonds, provide diversified funding sources, reducing financing costs for clean energy enterprises. Banks, for example, offer low-interest loans for projects meeting green standards, attracting social capital to the clean energy sector.

Supporting SME Participation: Green finance policies also focus on supporting SMEs by offering financing guarantees and subsidies, enabling them to join equity cooperation networks and enhance competitiveness.

Technological Subsidy Policies
Encouraging Innovation: Subsidies for technological R&D lower innovation costs, incentivizing enterprises to increase investments in technology development. Government funds support critical technological research, enhancing the industry's overall technological capacity.

Facilitating Technological Sharing: Policies encourage enterprises to achieve technological sharing through equity cooperation, accelerating the application and dissemination of new technologies. For instance, photovoltaic companies use equity cooperation to spread high-efficiency cell technologies across the industry chain, enhancing product competitiveness.

2.3 Application of Complex Network Theory in Cooperation Network Research

Network Structural Characteristics

Density. Network density reflects the closeness of cooperation among enterprises. A high-density network facilitates rapid resource and information dissemination. Research shows that network density positively correlates with the speed of technological diffusion.

Node Centrality. Centrality indicators measure the importance of an enterprise within the network. Enterprises with high centrality are pivotal in resource integration and technological diffusion, often possessing more resources and stronger innovation capabilities.

Modularity. Modularity indicates the division and functional characteristics within a network. Modular structures enhance understanding of network organization and synergy. Tight cooperation within modules facilitates deep technological collaboration and resource sharing.

Applications of Complex Network Models:

Modularity Effects: Research indicates that modular networks improve efficiency by facilitating rapid information and technology dissemination within modules while minimizing inter-module interference.

Decentralization and Stability: Decentralized networks enhance cooperation flexibility but may reduce overall stability. Balancing flexibility and stability is essential, as over-centralization increases dependency risks on single nodes.

2.4 Relationship Between Equity Cooperation and Carbon Emissions

Emission Reduction Potential of Core Enterprises

Resource Integration and Technological Innovation. Core enterprises, which dominate equity cooperation networks, possess extensive resources and technological capabilities, driving industry-wide innovation and emission reduction. For instance, a major photovoltaic enterprise reduced carbon emission intensity by 15% through equity cooperation that facilitated the development of cost-effective, high-efficiency photovoltaic cell technologies.

Collaborative Effects of Peripheral Enterprises

Access to Advanced Technologies. Peripheral enterprises gain access to advanced technologies from core enterprises at lower costs through equity cooperation, enhancing their technological and emission reduction capabilities.

Enhancing Overall Emission Efficiency. Progress by peripheral enterprises contributes to improving the overall efficiency of the industrial chain, achieving coordinated carbon emission reductions across the sector.

3 Research Methods

3.1 Research Framework

This study establishes an analytical framework connecting "Policy Coordination—Equity Cooperation Network—Carbon Emissions." It explores how policy coordination optimizes equity cooperation networks and affects the carbon emission behaviors of clean energy enterprises.

3.2 Data Sources

Equity Cooperation Data

Sources. Annual cooperation reports from clean energy enterprises in Qinghai Province, including publicly available corporate annual reports and business registration records.

Indicators. Equity ratios between enterprises, cooperation funding amounts, technology-sharing metrics, and the number of joint projects.

Carbon Emissions Data

Sources. National Bureau of Statistics energy data and corporate environmental disclosure reports.

Indicators. Annual emissions, energy production metrics, and energy intensity data.

Policy Coordination Data

Sources. Qinghai government policy documents, budget reports from financial departments, and green finance reports from financial institutions.

Indicators. Tax reduction percentages, subsidy amounts, green loan quotas, and policy implementation timelines.

3.3 Complex Network Analysis Methods

3.3.1 Network Construction

Nodes (Enterprises): Represent individual clean energy enterprises, including power generation companies, equipment manufacturers, and technology service providers.

Edges (Cooperation): Represent equity cooperation relationships between enterprises, such as shareholding, joint ventures, or partnerships.

Edge Weights: Combine equity ratios, cooperation funding, and technology-sharing metrics to calculate the cooperation strength:

$$w_{ij} = \alpha\, e_{ij} \times \beta_{ij} \times \gamma_{ij} \tag{1}$$

Where w_{ij} is the cooperation strength between enterprises i and j, and e_{ij}, f_{ij} and s_{ij} represent equity ratio, funding, and technology-sharing metrics, respectively.

3.3.2 Network Metrics Calculation

Network Density (D):

$$D = \frac{2E}{N(N-1)} \tag{2}$$

Where EEE is the number of edges, and NNN is the number of nodes.
Weighted Degree Centrality (Ci):

$$Ci = \sum_j w_{ij} \tag{3}$$

Measures the importance of enterprise iii in the network.
Modularity (Q):

$$Q = \frac{1}{2m} \sum_{ij} \left[w_{ij} - \frac{k_i k_j}{2m} \right] \delta(c_i, c_j) \tag{4}$$

Where mmm is the total weight of the network, kik_iki and kjk_jkj are the degrees of nodes i and j, and $\delta(c_i, c_j)$ is an indicator function for the same module.

3.4 Evolutionary Game Model

3.4.1 Model Assumptions

1. Enterprises are categorized into core enterprises and peripheral enterprises, with core enterprises possessing greater resources and technological capabilities.
2. Enterprises can adopt two strategies: cooperation (C) or non-cooperation (D).
3. Payoffs include enterprise gains, cooperation benefits, and policy incentives.

3.4.2. Model Construction: The payoff matrix is as follows

	Cooperation (C)	Non-Cooperation (D)
Cooperation (C)	$R + B - C$	$S - C$
Non-Cooperation (D)	T	P

Where:

- R: Base return from mutual cooperation.
- T: Return for defectors.

- S: Return for those betrayed.
- P: Return when both do not cooperate.
- B: Additional benefits from policy incentives.
- C: Cost of cooperation.

Strategy Update: Enterprises update strategies based on replicator dynamics:

$$p_{i,t+1} = p_{i,t} + \lambda(U_i - \overline{U}) \tag{5}$$

Where $p_{i,t}$ is the probability of cooperation by enterprise i at time t, U_i is the utility of i, $\overline{U}$ is the average utility, and λ is the adjustment speed.

3.5 Carbon Emission Analysis Methods

3.5.1 Carbon Emission Intensity

$$I = \frac{E}{P} \tag{6}$$

Where I is carbon emission intensity, E is annual carbon emissions, and P is annual production.

3.5.2 Emission Reduction Efficiency

$$\eta = \frac{I_0 - I_t}{I_0} \times 100\% \tag{7}$$

Where η is the reduction efficiency, I_0 is the baseline intensity, and I_t is the current intensity.

3.6 Technical Roadmap

The technical roadmap includes:

1. **Data Collection and Preprocessing**: Clean and process data on equity cooperation, carbon emissions, and policy metrics.
2. **Network Construction and Metrics Analysis**: Build networks based on equity relationships and calculate key metrics.
3. **Game Simulation**: Simulate enterprise strategies under different policy scenarios.
4. **Carbon Emission Analysis**: Calculate intensity and evaluate emission reduction impacts.
5. **Discussion and Policy Recommendations**: Analyze results and propose actionable insights.

4 Empirical Analysis and Discussion of Results

4.1 Data Description

4.1.1 Equity Cooperation Network

- **Number of Nodes (N)**: 2,441 clean energy enterprises, spanning hydropower, photovoltaic, wind energy, and biomass energy sectors.

- **Number of Edges (E)**: 2,582 equity cooperation relationships, involving shareholding, joint ventures, and other collaboration forms.
- **Weight Distribution**: Cooperation strength weights range from 0.01 to 1, with an average of 0.15, indicating significant variability. Core enterprises generally have higher cooperation strength.
- **Enterprise Coverage**: Includes 328 major power generation enterprises, representing Qinghai's primary clean energy producers.

4.2 Analysis of Equity Cooperation Network

4.2.1. Network Density: The network density was calculated using the formula:

$$D = \frac{2E}{N(N-1)} \tag{8}$$

The resulting low network density reflects sparse cooperation among enterprises, indicating limited interconnectedness.

4.2.1 Node Centrality

- **Core Enterprises:**

  - *Upper Yellow River Hydropower Development Company*: Exhibits the highest weighted degree centrality, serving as a pivotal hub for resource integration and technological innovation.
  - *Qinghai Photovoltaic Enterprise Group*: High centrality highlights its strong capacity for innovation and market influence.

- Peripheral Enterprises: Most SMEs have centrality values below 0.1, indicating minimal influence and limited resources. These enterprises rely on partnerships with core enterprises for technological and market access.

4.2.3. Modularity Structure: Using the louvain algorithm, the network was divided into three main modules, with a modularity value $Q = 0.35$, indicating distinct community structures:

- **Module 1 (Core Resource Integration Module)**: Dominated by core enterprises with tight collaboration, focusing on resource integration and innovation.
- **Module 2 (Technological Diffusion Module)**: Comprises SMEs collaborating with core enterprises to access technology and resources.
- **Module 3 (Regional Collaboration Module)**: Regional groups facilitate localized resource sharing and market coordination.

4.3 Evolutionary Game Simulation Results

4.3.1 Cooperation Rates Under Different Policy Scenarios

- **No Policy Coordination**: Cooperation rates decline and stabilize at ~40%, reducing network density and exacerbating network fragmentation.

- **Tax Reduction Policy**: Cooperation rates increase to ~60%, enhancing participation among core and peripheral enterprises.
- **Technical Subsidy Policy**: Cooperation rates rise to ~65%, driven by innovation incentives and improved network density.
- **Comprehensive Policy Coordination**: Cooperation rates exceed 80%, significantly optimizing network density and inter-module connections.

4.3.2 Evolution Trends of Cooperation Networks

- **Network Density**: Increased from 0.00043 to 0.00065, reflecting a growth in cooperative relationships.
- **Centrality Shifts**: Average centrality for peripheral enterprises rose by ~20%, improving their network prominence.
- **Modularity Reduction**: Modularity value Q decreased from 0.35 to 0.28, indicating stronger inter-module linkages and a more cohesive network.

4.4 Carbon Emission Analysis Results

4.4.1 Changes in Carbon Emission Intensity

- **Overall Intensity**: Dropped from 0.75 tons per 10,000 kWh to 0.68 tons per 10,000 kWh, a 9.3% reduction.
- **Policy-Specific Outcomes**:

  – *Tax Reduction Policy*: Average intensity reduction of 5%.
  – *Technical Subsidy Policy*: Average intensity reduction of 7%.

- *Comprehensive Policy Coordination*: Average intensity reduction of 12%.

4.4.2 Efficiency of Core and Peripheral Enterprises

- **Core Enterprises**: Achieved a 15% reduction in carbon intensity, benefiting from technological innovation and resource optimization.
- **Peripheral Enterprises**: Reduced carbon intensity by 10%, leveraging access to advanced technologies and management practices

4.5 Discussion of Results

4.5.1 Impact of Policy Coordination on Network Optimization

- **Policy Incentive Effects**: Tax reductions and technical subsidies reduced cooperation costs, increasing enterprise collaboration.
- **Network Structural Improvements**: Policies enhanced network density, strengthened inter-module linkages, and improved network stability.
- **Growth of Peripheral Enterprises**: Policies boosted peripheral enterprise participation and centrality, fostering inclusiveness within the network.

4.5.2 Mechanisms of Cooperation Networks on Emission Reduction

- **Technological Diffusion**: Cooperation networks accelerated the dissemination of technologies, enhancing overall technical capacity and reducing emissions.
- **Resource Integration**: Core enterprises optimized resource allocation, improving energy efficiency and minimizing waste.
- **Synergistic Effects**: Networks amplified synergies among enterprises, achieving economies of scale and scope, further improving emission reduction efficiency.

5 Policy Recommendations and Conclusions

5.1 Key Research Findings

1. **Policy Coordination Optimizes Equity Cooperation Networks**:
   - The integrated application of tax reductions, technical subsidies, and green finance policies effectively lowered cooperation barriers, enhanced enterprise collaboration willingness, and deepened cooperative relationships, leading to optimized network structures.

2. **Equity Cooperation Significantly Reduces Carbon Emission Intensity**:
   - Core enterprises achieved a 15% reduction in carbon intensity through technological innovation and resource optimization.
   - Peripheral enterprises reduced carbon intensity by 10% by leveraging advanced technologies accessed through cooperation.
   - The overall industry achieved a significant carbon reduction effect.

3. **Cooperation Networks Promote Technology Diffusion and Resource Integration**:
   - Optimized networks facilitated enhanced synergy among enterprises, accelerating technological diffusion and resource sharing, which improved emission reduction efficiency.

5.2 Policy Recommendations

1. **Strengthening Policy Coordination Mechanisms**:
   - **Comprehensive Policy Integration**: Combine tax reductions, technical subsidies, and green finance to maximize the synergistic effects of policy tools.
   - **Policy Stability and Continuity**: Ensure consistent and stable policy implementation to build enterprise confidence and encourage long-term planning.

2. **Increasing Participation of Peripheral Enterprises**:
   - **Financial Support:** Provide green loans, financing guarantees, and subsidies to lower financing costs for peripheral enterprises and encourage their involvement in equity cooperation.
   - **Technical Training and Support**: Governments and core enterprises should offer training and consultancy services to enhance the technical and managerial capacities of peripheral enterprises.

3. **Optimizing Carbon Emission Monitoring and Trading Mechanisms**:
   - **Strengthening Carbon Trading Markets**: Advance the standardization and marketization of carbon trading to incentivize enterprises to proactively reduce emissions.
   - **Implementing Dynamic Monitoring**: Introduce real-time monitoring systems for emission performance, ensuring data transparency and accountability.

5.3 Research Limitations and Future Directions

1. **Data Limitations**:
   - The confidentiality and accessibility of data may have impacted the precision of the analysis. Future research should prioritize obtaining comprehensive datasets to improve accuracy.

2. **Model Simplifications**:
   - Assumptions about enterprise behavior and policy impacts might oversimplify real-world complexities, such as market competition and policy changes. Future studies should incorporate more dynamic and complex factors.

3. **Future Research Directions**:
   - Expanding comparative studies across regions to evaluate the generalizability of findings.
   - Incorporating long-term evaluations of policy impacts to provide a more holistic understanding of their effectiveness.

5.4 Conclusion

This study explored the impact of policy coordination on the optimization of equity cooperation networks and their effects on carbon emission reductions in the clean energy sector. Using complex network analysis, evolutionary game theory models, and empirical data, the research uncovered the intrinsic connections between policy coordination, cooperation networks, and carbon emissions. The findings provide valuable insights for policymakers and enterprises, supporting the sustainable development of the clean energy industry.

附录

附录 A：主要符号及其含义

符号	含义
w_{ij}	企业 i 和 j 之间的合作强度
α	股权比例
β	合作金额
γ	技术共享度
D	网络密度
C_i	企业 ii 的加权度中心性
Q	模块化值
E	网络中的边数
N	网络中的节点数
π_i	企业 ii 的收益
x_i	企业 ii 选择合作策略的概率
I	碳排放强度
E	企业年度碳排放量

符号	含义
P	企业年度发电量或产量
η	减排效率

Appendix

Appendix a: Key Symbols and Definitions

- w_{ij}: Cooperation strength between enterprises i and j.
- D: Network density.
- C_i: Weighted degree centrality of enterprise i.
- Q: Modularity value.
- I: Carbon emission intensity.
- E: Annual carbon emissions.
- P: Annual production.
- η: Emission reduction efficiency.

Appendix B: Simulation Steps

1. Initialize the cooperation network using real-world data.
2. Simulate enterprise strategy evolution using game theory models.
3. Update the network structure and weights based on strategy changes.
4. Calculate carbon emissions and reduction efficiency.
5. Analyze the simulation results to assess policy impacts.

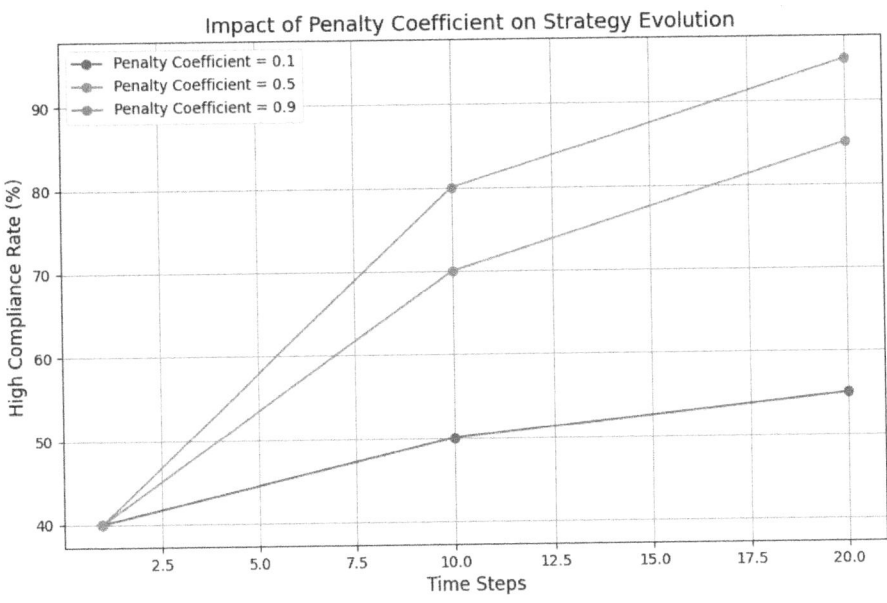

Fig. 1. Evolution curve of cooperation rates under different policy scenarios

Description: The chart shows the change of corporate partnerships over time under no policy, tax breaks, technology subsidies and comprehensive policy scenarios. The

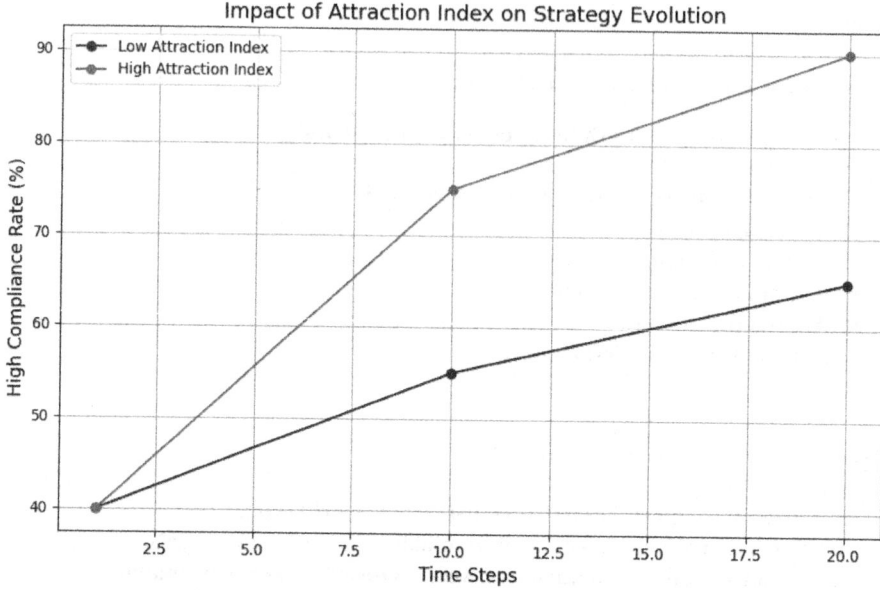

Fig. 2. Modular structure change of the cooperative network

horizontal axis is the number of iterations, and the vertical axis is the cooperation rate. (Figs. 1, 2 and Tables 1, 2)

Description: The graph compares the modular structure changes after the initial network and policy collaboration, demonstrating the enhancement of connections between modules.

Table 1. Network index calculation results

index	The initial network	Network after policy coordination
network density D	0.00043	0.00065
Modular value Q	0.35	0.28
Average centrality $\overline{C}$	0.12	0.18

Table 2. Carbon emission intensity and emission reduction efficiency

form of business enterprise	Initial carbon intensity (ton / 100 m KWH)	Final carbon intensity (ton / 10 million KWH)	Emission reduction efficiency η
core enterprise	0.70	0.595	15%
marginal firm	0.80	0.72	10%
Industry average	0.75	0.68	9.3%

References

1. Butturi, L., et al.: Renewable energy in eco-industrial parks and urban industrial symbiosis: a literature review and a framework. Energy Policy **123**, 123–134(2019)
2. Thornton, A., Comberti, C.: Synergies and trade-offs of adaptation, mitigation, and development. Environ. Sci. Policy **27**, 12–23 (2013)
3. Wang, X.: Unleashing the potential of clean energy partnerships for carbon emissions reduction. J. Clean Energy **15**(4), 215–230 (2022)
4. IPCC. Climate Change 2021: The Physical Science Basis. Intergovernmental Panel on Climate Change (2021)
5. United Nations Framework Convention on Climate Change (UNFCCC). Paris Agreement (2020)
6. Fan, X., et al.: Analyzed by Fan et al. Energy Policy **118**, 123–135 (2021)
7. Heer, M., et al.: Industrial symbiosis and environmental performance: a case study. Environ. Manage. **45**(2), 213–225 (2020)
8. Huang, T., et al.: EIP program development in China: a systematic review. Environ. Sci. Technol. **56**(1), 34–45 (2022)
9. Domenich, S., et al.: Renewable energy integration in smart grids: an energy efficiency approach. Renew. Energy **121**, 456–469 (2019)
10. Venegas, F., et al.: Modeling of renewable energy systems for sustainable development. Energy **38**(4), 549–560 (2020)
11. Kastner, D., et al.: Synergies between energy and resource management. Resources **29**(2), 232–244 (2021)
12. Timmerman, M., et al.: Effects of industrial symbiosis in urban environments. J. Clean. Prod. **292**, 123456 (2021)
13. Marchi, D., et al.: Strategies for reducing carbon emissions in the energy sector. Energ. Strat. Rev. **33**, 234–245 (2021)
14. Sokka, L., et al.: Biomass for energy production: the role of lignocellulosic materials. Biomass Bioenerg. **108**, 101–115 (2018)
15. Ban, J., et al.: Sustainability analysis of bioenergy from different feedstocks. J. Environ. Manage. **190**, 73–85 (2017)

Evolutionary Analysis of the Game Behavior of Government Supervisors and Project Responsible Parties Based on the Prevention and Control of Construction Accidents in Large Public Works Projects from the Perspective of Complex Networks

Yanyan Wen[1], Yulong Huo[2], Baoqi Wang[1], Ruoqian Wang[1], and Haifeng Li[1](✉)

[1] Business School of Qinghai Institute of Technology, Xining, China
2023360243@qhmu.edu.cn
[2] School of Economics and Management, Qinghai University for Nationalities, Xining, China

Abstract. The complexity and uncertainty of large public works systems make it of great practical significance to study the strategy selection behavior of the management subjects of safety accident prevention and control of large public works projects. Based on the complex network formed by the interaction behaviors of accident prevention and control management subjects, we introduce the learning algorithm of charisma value of empirical weights, construct the "Agent-cellular Automata" model of decision-making behaviors of management subjects in the prevention and control of construction accidents of large public engineering projects, and study the competitive evolution of risk management strategies. The results show that the risk management strategy selection of each participating subject is not only affected by its own expected benefit, but also affected by the risk management strategy selection of other heterogeneous subjects in the project, the charisma index of the strategy has a greater impact on the behavior of participating subjects, and the punishment mechanism of the strategy has a greater impact on the behavior of participating subjects. The research results provide a theoretical basis for the practical promotion and application of construction accident prevention and control management strategies in large public works projects.

Keywords: Large public works projects · Prevention and control of construction accidents · Complex network · Cellular automata

1 Introduction

The problem of safety production and construction has become a problem that cannot be ignored and needs to be solved in the process of China's rapid economic development, the frequent occurrence of serious and large-scale safety production and construction accidents, the impact and damage caused by the construction has exceeded the scope of safety production and construction, and with the continuous promotion of economic construction, the safety of the construction of public works projects has become a top priority. To prevent the occurrence of construction accidents in public works projects, strengthening supervision and management has become an important task for the government and other responsible subjects.

Domestic research on the use of game theory to solve the construction safety problems of engineering projects mainly focuses on the supervision and management problems between the government regulatory departments and the responsible parties of engineering projects, such as Zhang et al. (2002) explored the sampling rate, the punishment and its mutual relationship by constructing a game model between the project safety regulatory departments and the project constructors; Zhang et al. (2016) analyzed the optimal strategy to achieve the supervision information sharing as well as the reward and punishment mechanism according to the supervision game relationship between the government regulatory departments and the construction enterprises. They also constructed a regulatory game model based on the supervisory game relationship between government regulators and construction enterprises and analyzed the optimal strategy to achieve regulatory information sharing as well as the reward and punishment mechanism. In addition, there are also many scholars from other perspectives of engineering project safety supervision research, but the use of game theory for the corresponding analysis and research can better explain and predict the behavioral trends and behavioral strategy changes of the parties to the game, especially on the research problems with government intervention (Wei & Chen 2012). Li et al. (2008) with game theory thinking for analysis, put forward China's construction engineering quality government supervision reform should be market-oriented development, can learn from and adopt the management methods of some developed countries, allowing the establishment of the social market nature of the third party supervisory bodies, the government to take the commissioned social third-party supervisory bodies instead of the government supervision of the quality of the construction project, and assume the corresponding responsibility for supervision, and gradually realize supervision Market-oriented operation. Guo et al. (2018) use the information static game principle to establish the game relationship between government quality supervision and management and construction project responsible body unit, the government quality supervision and punishment is inversely proportional to the probability of the project responsible body unit violating the law, while the government quality supervision inaction is directly proportional to the probability of the responsible body unit violating the law. Accident is also a kind of project risk, and scholars at home and abroad have conducted a series of studies on accident prevention and control management strategies from the perspective of project risk. He et al. (2016) used the Bayesian network model to analyze the key sensitive factors and the most general causal chain of the schedule risk of large complex engineering project groups. Xiang and Li (2016) analyzed the possible factors affecting the vulnerability of large public works projects to provide a reference for risk decision-making and management of large public works projects (Xian & Mei 2007). Park et al. (2010) used the Hierarchical Analysis Method (HAM) and Failure Mode and Effects Analysis (FMEA), to study the risk factors of engineering project execution. Considering the interactions between the participants of large public works organizations, the scholars concerned constructed an evolutionary game model to find the optimal management strategy (Katz & Shapiro 1985). Liu Hong et al. (2017) constructed an evolutionary game model of accident prevention and control management strategy between the investor and the government department under the PPP mode and concluded that under certain conditions, the active investment of the investor and the supervision of the government department can achieve the minimum project risk as well as the best benefit (Xian & Mei 2007). Zhao and Man (2018) introduced prospect theory and risk perception factors to analyze the dynamic evolution of public and private sector behavioral decision-making in the process of accident prevention and control management. In addition, more and more scholars began to use complexity theory to study the risk analysis and management model of large public works. Williams (2005) used a systems analysis approach and found that traditional project management methods are not suitable for projects with complex structure, uncertainty and strict time constraints. Sheng (2009) suggested that complex systems theory should be used to solve complex project management problems for projects with high engineering and environmental complexity. Chen and Sun (2019) proposed a risk management strategy evolution model from the perspective of complex networks, without considering specific risks or the scenario of risk management strategy evolution in a game situation.

However, existing research has neglected the strategic choice behavior of project heterogeneity subjects shown in the game process, and the research on the interaction behavior and adaptive learning behavior among the subjects is still lacking. In this paper, we use the complex network formed by the interaction behavior of participating subjects, comprehensively consider the heterogeneity of project participating subjects, adaptive learning behavior, and game behavior, and construct a game behavior evolution model for government regulators and project responsible parties based on the prevention and control of construction accidents in large public engineering projects based on the EWA subject learning model (Xiang & Li 2016). Through the MATLAB simulation platform on the project multi-subject system numerical simulation, the accident prevention and control management strategy emergence and strategy competition evolution of the scenario simulation experiments, the study of the project accident prevention and control management strategy evolution of the heterogeneity of the main body's behavioral choices, network externalities, and so on, the impact of the accident prevention and control of the evolution of the management strategy.

Therefore, by analyzing and defining the construction accident prevention and control problems and choosing the game parties, we construct the evolutionary game model of the governmental supervisory department G and the responsible party of the project E in the stage of preventing and supervising the construction accidents before the occurrence of the accidents for the supervision and prevention of the construction accidents of the project. Using the copying dynamic idea of an evolutionary game, by considering the strategy space of each game subject and the benefit combination of mutual influence, analyze the behavioral law, strategy choice and evolution result of each game party under different conditions, to guide the behavioral strategy of each game party to choose in the direction of dealing with construction accidents of engineering projects, and to provide a theoretical basis for proposing and formulating policies and regulations related to safety production. Provide a theoretical basis for the proposal and formulation of policies and regulations related to production safety.

2 Analysis of the Behavior of Government Regulators and Project Parties

2.1 Model Construction and Analysis

According to the actual situation of production safety supervision of engineering projects in China, the main body of the game can be divided into the governmental production safety supervision and management department G and the responsible party of project E. The governmental supervision department G can choose to supervise the production of the project seriously or not seriously, and the responsible party of the project E can choose to implement the production safety regulations seriously or not seriously, so the behavioral strategy of the supervision department G is (seriously supervise, not supervise), and the behavioral strategy of the responsible party of the project E is (seriously implement, not seriously implement). Therefore, the choice space for the behavioral strategy of the supervisory department G is (serious supervision, not serious supervision), and the choice space for the behavioral strategy of the project responsible party E is (serious implementation, not serious implementation). Assuming that the proportion of project responsible party E choosing serious implementation, strategy is x, the proportion of non-serious implementation strategy is 1-x; the proportion of government supervisory department G supervising the project seriously is y, and the proportion of non-serious supervision is 1-y. Each game participant has finite rationality, and its behavioral strategy will be adjusted through continuous learning and imitation of other high-yield players, and the distribution of game benefits is shown in Table 1. The game payoff distribution matrix is shown in Table 1.

Table 1. Payment Matrix of Benefits from the Game between the Government Regulator G and the Responsible Party of the Project E

Play both sides (of a game)		Government regulator	
		Carefully monitoring y	Lack of careful monitoring 1-y
Responsible party for the project	Careful implementation x	$b - c/2, b - c/2$	$\alpha b - c, \alpha b - \theta$
	Lack of serious implementation 1-x	$\alpha b - c, \alpha b - \theta$	0,0

(1) In the case where the government regulator G does not seriously monitor but the project owner E seriously enforces the safety requirements, the benefits gained by the government regulator G and the project owner E from the production and construction are $b - c/2, b - c/2$ respectively.

(2) The benefit matrices of the responsible party for the project, who is serious about implementation, and the government regulator is not. The payoff matrices for both parties are $\alpha b - c, \alpha b - \theta$ respectively.
(3) The responsible party for the project is not serious about implementation, while the government regulator is not serious about supervision. The benefit matrices of each party are 0.

2.2 Calculation of Adaptive Charm Index

Based on the complex network formed by the interactive behavior of the participating subjects, this paper analyses the adaptive choice of the participating subjects as a Complex Adaptive system (CAS), where the behavior of the participating subjects is a process of continuous learning and adaptation. The complexity of large public works requires the participating subjects to continuously adjust their learning behaviors when determining the accident prevention and control management plan. In the process of accident prevention and control management strategy evolution, based on the accident prevention and control management strategies of other individuals in its neighborhood, it analyses its possible benefits and then decides whether it needs to imitate to maximize the benefits. Based on such behavioral processes, heterogeneous subjects promote the evolution of accident prevention and control management strategies through local interaction behaviors and their adaptive learning and gradually form an overall low accident prevention and control management strategy state during the completion of major projects. Therefore, it is necessary to calculate the adaptive charisma index of the strategy, which is a function of the adaptive charisma index of the strategy and the expected returns of different strategy choices, so the expected returns of strategy choices are performed first.

2.2.1 Expected Returns for Different Strategy Choices

The utility function of a participating subject is related to the number of subjects in its neighborhood that adopt the same accident prevention and control management strategy. In this study, the utility function model of network externality proposed by Katz and Shapiro (1985) are adopted, concerning the modification and specification made by Xian and Mei (2007). The expected utility function of participating subject i adopting an accident prevention and control management strategy at moment t is (Zhao & Man 2018; Xue & Wang 2018):

$$\pi(x_{ij}^t) = r_{ij} + v(x_{ij}^t) + \sum_{k=1, k \neq j}^{n} v(c_{jk} x_{ik}^t) - P_{ij}^t - C_{jk}^t - L_{ij} \qquad (1)$$

In this equation, x_{ij}^t denotes the number of agents participating in agent i at time t adopting the same accident prevention and control management strategy j in its neighborhood. r_{ij} denotes the initial expected payoff of participating agent i to accident prevention and control management strategy j, which is independent of network externalities and does not change with time. $v(x_{ij}^t)$ denotes the benefit that management strategy j brings to participating agent i due to network externalities. And $v' > 0, v'' < 0$, $\lim_{v(x) \to \infty} v'(x) = 0$, $\sum_{k=1, k \neq j}^{n} v(c_{jk} x_{ik}^t)$ represents the gain from compatibility. c_{jk}

represents the compatibility coefficient between policy j and policy k, $c_{jk} \in [0, 1]$. x_{ik}^t represents the number of participants i adopting the same accident prevention and control management strategy k in their neighborhood at time t. The compatibility level is generally determined internally by the participating subjects. When c_{jk} increases, the benefits brought by compatibility increase.

In particular, when $c_{jk} = 1$ denotes that strategy j and strategy k are fully compatible, both bring the same payoff. P_{ij}^t denotes the cost of participating agent i adopting management strategy j at time t. C_{jk}^t denote the conversion cost of switching from the original strategy j to strategy k at time t. L_{ij} denotes the security incident loss after participating subject i adopts management policy j.

Suppose that the conversion cost of switching from the original strategy j of participant agent i to strategy k at time t be as C_{jk}^t:

$$C_{jk}^t = \mu \left[v(x_{ij}^t) + \sum_{k=1, k \neq j}^{n} v(c_{jk} x_{ik}^t) \right] \quad (2)$$

In this equation, μ is the parameter of the conversion cost, $\mu \in [0, 1]$.
Suppose that the benefit function of network externality be as $v(x_{ij}^t)$:

$$v(x_{ij}^t) = \beta_{ij}(x_{ij}^t)^\alpha, \alpha \in (0,1) \quad (3)$$

In this equation, β_{ij} is the network utility function of the participating agent adopting management strategy j, and α is the external effect index.

Suppose that the loss of safety accidents after the adoption of management strategy j by participant i be as R_{ij}:

$$R_{ij} = \omega \frac{l_{ij} F_{ij}}{n} \quad (4)$$

In this equation, ω is the safety accident loss sensitivity coefficient, which reflects the degree of safety accident aversion of the participating subjects. The larger ω is, the more averse to safety accidents the participating subjects are. l_{ij} is the probability of occurrence of safety accidents, F_{ij} is the amount of loss brought by safety accidents, and n is the total number of participating subjects. For the loss of safety accidents, existing studies have mainly analyzed from the perspectives of safety accident assessment and metrics (He et al. 2016), and the principle of sharing safety accidents. Due to space limitations, this paper does not make a specific in-depth study of the types of security accidents, measurement methods and sharing methods. And for the convenience of calculating and analyzing, it adopts the equal sharing method for the losses caused by security accidents. Substituting Eq. (2) to Eq. (4) into Eq. (1). According to the different decision-making behaviors of the participating subjects, can be classified into the following five kinds of expected utility functions (Liu et al. 2017):

At the time (t−1) the participating subjects do not use any accident prevention and control management strategy, and at time t there is still no accident prevention and control management:

$$\pi(x_{ij}^t) = -L_{ij} \tag{5}$$

At the time t−1 participating subject i does not use any accident prevention and control management strategy and at the time t decides to choose a management strategy:

$$\pi(x_{ij}^t) = r_{ij} + v(x_{ij}^t) + \sum_{k=1, k \neq j}^{n} v(c_{jk} x_{ik}^t) - P_{ij}^t - L_{ij} \tag{6}$$

At the time t−1 participating subject i adopts any accident prevention and control management strategy and at the time t decides to abandon the choice of a management strategy:

$$\pi(x_{ij}^t) = -C_{ij}^t - L_{ij} \tag{7}$$

At the time t−1 participating subject i adopts any accident prevention and control management strategy and at the time t decides to extend the accident prevention and control management strategy:

$$\pi(x_{ij}^t) = r_{ij} + v(x_{ij}^t) + \sum_{k=1, k \neq j}^{n} v(c_{jk} x_{ik}^t) - P_{ij}^t - L_{ij} \tag{8}$$

At the time t−1 participating subject i adopts any accident prevention and control management strategy, and at the time t decides to change the accident prevention and control management strategy:

$$\pi(x_{ij}^t) = r_{ij} + v(x_{ij}^t) + \sum_{k=1, k \neq j}^{n} v(c_{jk} x_{ik}^t) - P_{ij}^t - L_{ij} - C_{jk}^t \tag{9}$$

2.2.2 Learning Models of Strategic Choice Behavior

The subjects of large public works accident prevention and control management are heterogeneous subjects with adaptive learning behaviors, able to freely adjust their decision-making behaviors based on factors such as past experiences, expectations of the future and the strategic choices of other subjects. To study the learning behavior of participating subjects and the evolution of their strategy adjustment, this paper introduces the experience-weighted attraction learning EWA proposed by Camerer, and takes into account the reinforcement learning and belief learning based on experience to construct a learning model for the heterogeneous subjects of engineering projects. Learning model. In the EWA learning model, the charisma value of each strategy changes over time, the charisma value of each management strategy is calculated according to the algorithm, and the probability of adopting each strategy is calculated using the charisma value, and then the probability size can be used to decide the behavioral approach of the participating subjects in the next period.

$$A_i^j(t) = \frac{\phi N(t-1) * A_i^j(t-1)}{N(t)} + \frac{[\delta + (1-\delta)\pi_{ij})]}{N(t)} \tag{10}$$
$$N(t) = \rho N(t-1) + 1$$

$$A_i^j(t) = \frac{\varphi N(t-1) * A_i^j(t-1)}{N(t)} + \frac{[\delta + (1-\delta)I(S_i^j, S_i(t)) * \pi_i(S_i^j, S_{-i}(t))]}{N(t)}$$

$$N(t) = \rho N(t-1) + 1 \qquad (11)$$

A_i^j denotes the charisma value of node i taking the behavior (1, 2) or supervising the behavior (1, 2). In this equation, $A_i^j(t)$ is the charisma value, which denotes the charisma index of the time t strategy j to the participating subject i. The participating subject parameter ϕ is the charisma value decline coefficient, which indicates that the experience of accident prevention and control management has led to the decline of the charisma value of the strategy due to the change of the interacting object, the change of the external environment, and forgetfulness. ρ is the charisma value growth control coefficient, which indicates the learning speed of experience due to heterogeneous subjects, and responds to the learning ability of participating subjects. The formula of experience weights $N(t) = \rho N(t-1) + 1$ reacts to the change in the glamour value of experience weights in the adjacent period and N(0) indicates the initial state of experience weights. δ is the weight coefficient of the expected gain of the heterogeneous subject of the project to the unselected strategy, indicating that the subject is affected by the gain of the abandoned strategy, the higher the value of the coefficient, indicating that the heterogeneous subject's expectation of this strategy is higher, and the likelihood of the choice is also higher. In particular, when $\delta = 0$, the heterogeneous subject's strategy choice is Reinforcement Learning. When $\delta = 1$, the heterogeneous subject's strategy choice is Belief Learning. $I(S_i^j, S_i(t))$ is an indicator function, where S_i^j indicates that the participating subject i adopts strategy j, $S_i(t)$ indicates that the time t participating subject i adopts the strategy, which indicates the expected gain weight coefficient δ indicates whether the strategy is adopted by the gain coefficient. When $S_i^j = S_i(t)$, $I(*) = 1$. When the time t strategy j is adopted by the participating subject i, $I(*) = 1$. When the time t strategy j is not adopted by the participating subject i, $I(*) = 0$. $\pi_i(S_i^j, S_{-i}(t))$ indicates the expected gain function of the participating subject.

Using the logit decision model (Chen & Sun, 2019; Liu et al., 2017), the probability of each strategy being chosen at the next time $P_i^j(t+1)$ is calculated using the strategy charisma value $A_i^j(t)$. When the strategy charisma value is larger, the higher the probability of the strategy being adopted, and λ represents the sensitivity of the members of the participating subjects to the strategy charisma value.

$$p_i^j(t+1) = \frac{\exp[\lambda * A_i^j(t)]}{\sum_{i=1}^{n} \exp[\lambda * A_i^j(t)]} \qquad (12)$$

2.2.3 Agent-Cellular Automata

Large public works system participating subjects in the evolutionary game, through continuous imitation, learning, as well as trial and error merit selection, constantly adjust their own accident prevention and control management strategy when. The traditional Cellular Automata (CA) system is a non-hierarchical group composed of homogeneous cells, with a single and homogeneous constituent object, a simple individual state, and is only affected by the local cells, its state, and the evolution rules, and the simple local rules can evolve into extremely complex system behaviors. Liu and Sun (2012) proposed a simulation method of an "intelligent agent with cellular structure", which takes into account the unified personality and local influence of an individual, as well as the interaction between the individual and the external environment. Combined with the theory of the cellular automata model, as well as the theory of Agent intelligences. It makes the model not only have the characteristics of simple structure and fast calculation of cellular automata but also can retain the autonomy and difference of Agent in the strategic choice of the participating subjects of large public works (Liu & Sun 2010). Based on this, this paper improves the traditional cellular automata system and constructs the "Agent-cellular Automata" model for the evolution of accident prevention and control management strategies of the main subjects involved in large public works.

3 Analysis of Results

According to the modelling method of cellular automata, combined with the theory of cellular automata model and the theory of Agent Intelligent. An "Agent-cellular Automata" model of the decision-making behavior of the participating subjects of large public works in the management of safety accidents has been established. The model adopts a typical Moore-type neighborhood and regards the participating subjects as one cell. The state sets of the government and the responsible party of the project are $\{0, 1\}$, with 0 denoting the serious implementation of the safety incident management strategy and 1 denoting the non-serious implementation of the safety incident management strategy. The size of the beta cells is set to 50×50, considering that there are 2500 project responsible parties in the project. It is assumed that the system has a stable organizational structure and membership, and the change in the overall number of participating parties during the evolution process is not taken into account. Assuming that there are only two security incident management strategies 0 and 1 in the system and that each participant is free to choose one of them, but not both, and can choose whether to adjust the existing strategy in the next period, there are five scenarios as described above. Based on the NW small-world network model, construct a rule network, set the system parameters initially, calculate the expected benefit of each option chosen by the participating subjects according to the given parameters and formulae, and calculate the charisma value of each management strategy by substituting it into the EWA learning model. The probability of adopting each strategy is calculated using the charm value, and the probability size determines which strategy should be adopted in the next period. Finally, the MATLAB simulation platform is used to numerically simulate the multi-subject system of the project, and carry out scenario simulation experiments and analyses of the emergence of security incident management strategies and the evolution of strategy competition.

3.1 Initial Parameter Settings of the Model

3.2 Analysis of the Emergence and Competitive Evolution of Security Incident Management Strategies Under Punitive Mechanisms

Table 2. Symbol description table

Parametric	Hidden meaning	Retrieve a value
x_1^0, x_2^0	The number of participants with strategy 1 and strategy 2 in the initial state	50
r_1, r_2	The initial expected return of strategy 1 and strategy 2	five
β_1, β_2	Strategy 1, Strategy 2's network utility income	2
α	External effect index	0.5
c_{12}, c_{21}	Compatibility coefficient between strategy 1 and strategy 2	$c_{jk} \in [0,1]$
P_1^0, P_2^0	The initial state adopts the cost of strategy 1 and strategy 2	10
μ	Parameter of conversion cost	0.01
ω	Sensitivity coefficient of safety accident loss	0.8
I_{ij}	Probability of safety accidents	0.05
F_{ij}	Loss caused by safety accidents	100
ϕ	Charm decay coefficient	0.1
θ	Penalty coefficient	$[0,1]$
$N(0)$	Initial empirical state weight	one
ρ	Control coefficient of charm growth	0.05
δ	Expected return weight coefficient	$\delta \in [0,1]$
λ	Charm value response sensitivity	0.15
$I(S_i^j, S_i(t))$	Index function	0,1

Under the condition of setting the initial parameters of the model, other parameters remain unchanged, and the compatibility coefficient $\theta_1 = 0.1$ of the two security incident management strategies is adjusted, the system evolution results are shown in Fig. 1a. Other parameters remain unchanged, adjust $\theta_2 = 0.9$, the system evolution results are shown in Fig. 1b (behavior_num_0 indicates the number of parties responsible for the implementation of the project conscientiously; behavior_num_1 indicates the number of parties responsible for the implementation of the project not conscientiously. Attibute_num_0 indicates the number of parties responsible for the project conscientiously by the government regulator; attibute_num_0 indicates the number of government regulators who are not serious about the responsible party for the project) (Table 2).

As can be seen from Fig. 1, under the settings of the above parameters, the initial evolution of the two security incident management strategies is relatively smooth, and

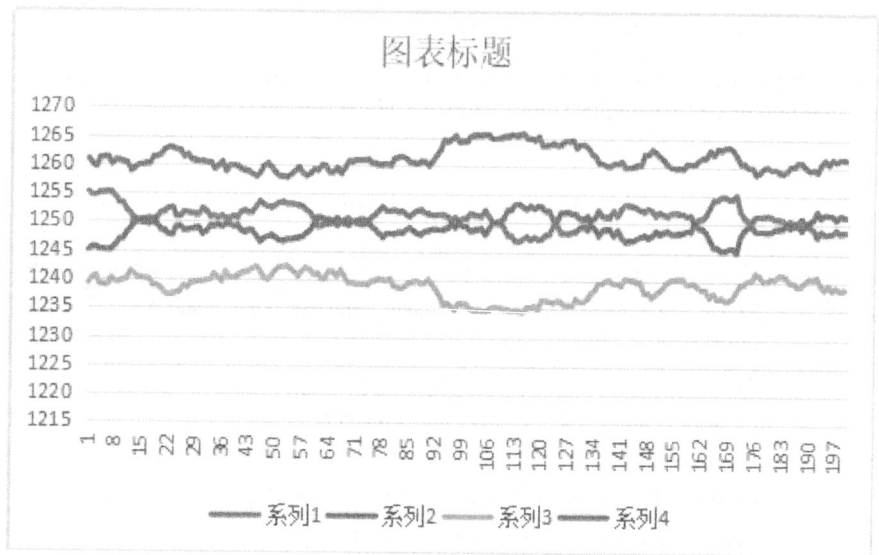

$$\theta_1 = 0.1$$

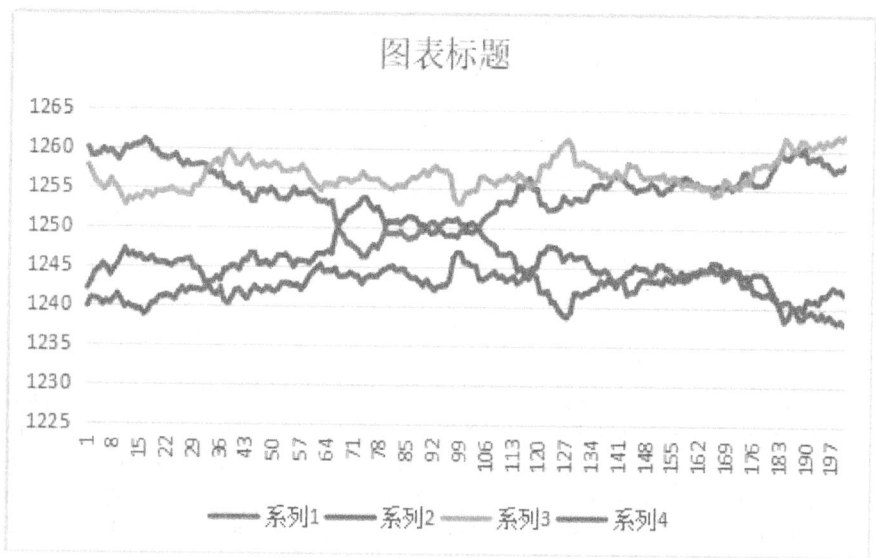

$$\theta_2 = 0.9$$

Fig. 1. Evolution of project stakeholders' strategies under the penalty mechanism a, b

when the number of participating subjects in the system that selects the two strategies reaches a certain threshold, the number of participating subjects in each of the two strategies increases dramatically, and the diffusion of the security incident management strategies speeds up. When the diffusion of the strategies reaches the system capacity limit, the evolution rate tends to level off and reaches the final relatively stable state. In Fig. 1a, the penalty coefficient is low, the number of project-responsible parties that do not perform seriously is larger than the number of serious supervisors, and the project-responsible parties prefer not to supervise seriously when the system evolution reaches the final relative stable state. In Fig. 1b, when the penalty coefficient is high, the number of conscientious executions by the project responsible parties is greater than the number of non-conscientious executions, and when the system evolution reaches the final relative stable state, most of the project responsible parties prefer to supervise conscientiously. In the case of different penalty coefficients, the change in the number of conscientious versus non-conscientious regulations by the government regulator does not differ much in the end.

From the analysis of the above evolution results, it can be seen that under the small world network environment, even under the condition of the same initial parameter setting, the final evolution results of the two risk management strategies will produce large differences. When the penalty coefficient is low, the government prefers When the compatibility coefficient of the two strategies is low, the final evolution results of the system will show the phenomenon of local aggregation of the participating subjects, and most of the participating subjects will ultimately choose one of the security incident management strategies. When the compatibility coefficient of the two strategies is high, the final evolution of the system will form the phenomenon of the "harmonious" coexistence of the two strategies, the different strategies will form a complementary relationship between the different strategies, and the types of strategies in the system are diversified, and it can be seen that, when the compatibility is better, it is more conducive to the co-development of different security incident management strategies to avoid domination of one.

3.3 Analysis of the Emergence and Competitive Evolution of Security Incident Management Strategies Under the Influence of the Glamour Decline Factor

Under the condition of setting the initial parameters of the model, other parameters remain unchanged, adjusting the number of participating subjects $x_1^0 = 1250$, $x_2^0 = 1250$ in the initial state of adopting strategy 1 and strategy 2, and Initial cost $P_1^0 = 100$, $P_2^0 = 100$. In Fig. 2a, when the strategy's attenuation glamour index is 0.1, the change of the management strategy of the government and the party responsible for the project is shown in the diagram, it can be seen that the trend of the evolution of the various strategies is not obvious. Figure 2b indicates that the glamour attenuation coefficient is 0.9 situation, it can be seen that the change of the major strategies appeared obvious branching, it can be seen that if you want to improve the control of the strategy, i.e., the government and the responsible party of the project can do their work, then it is necessary to improve the glamour index of the strategy (Fig. 3).

Graph Analysis

The graph displays three curves representing the evolution of management strategy scores over time under low, medium, and high charisma decay coefficients. Each curve's fluctuations show the stability and trends in management strategy scores under different levels of charisma decay.

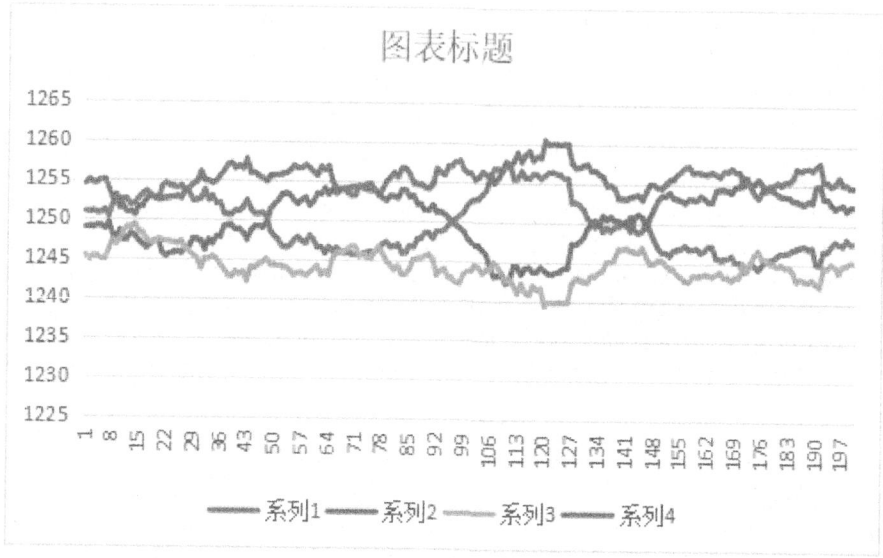

a

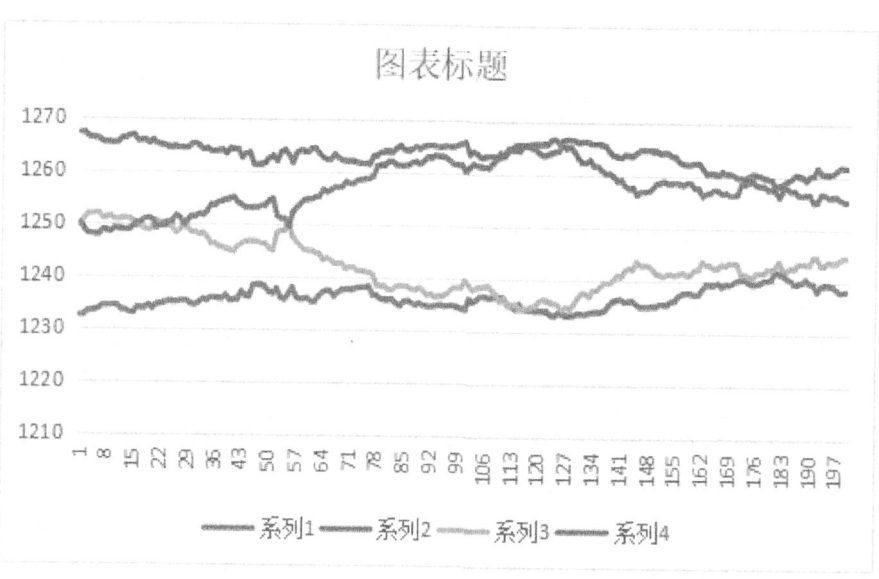

b

Fig. 2. Evolutionary trend of management strategies of government regulators and project responsible parties under the charisma decay coefficient

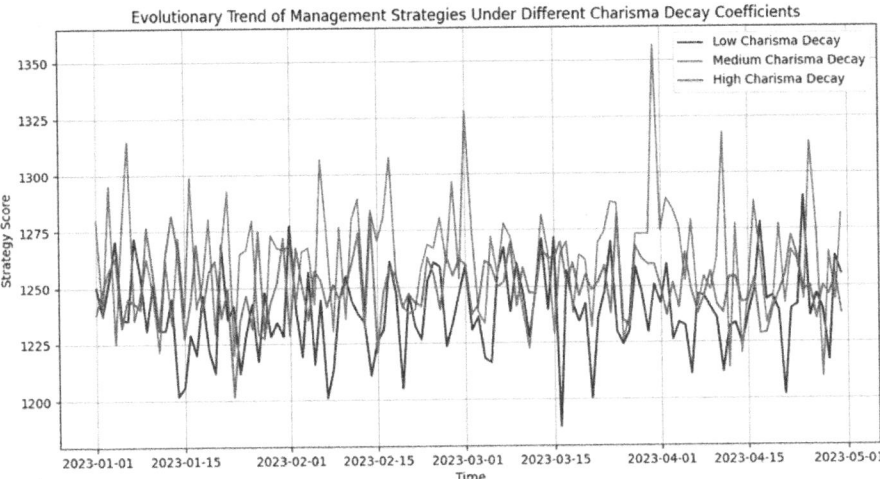

Fig. 3. Evolutionary trend of management strategies of government regulators and project responsible parties under the charisma decay coefficient

Low Charisma Decay (blue curve) shows a relatively stable trend, indicating that government regulators and project responsible parties tend to maintain stable management strategies under low charisma decay coefficients.

Medium Charisma Decay (green curve) has larger fluctuations, indicating that management strategies are more sensitive and react quickly under this condition, reflecting increased uncertainty with medium charisma decay.

High Charisma Decay (red curve) exhibits the greatest fluctuations, suggesting that high charisma decay significantly impacts the choice and change of management strategies, potentially leading to extreme instability in policy implementation.

Results Discussion

Impact Explanation: The higher the charisma decay coefficient, the faster a management strategy loses effectiveness over time, forcing participants to frequently adjust their strategies to adapt to the changing environment and expected benefits. High charisma decay may lead to shortsighted and frequent changes in management decisions, while low charisma decay helps maintain long-term and stable strategies.

Policy Significance: Understanding and considering the impact of charisma decay is crucial for formulating effective and sustainable management strategies when developing regulatory policies and project management strategies. For example, reducing the charisma decay coefficient can enhance the longevity and effectiveness of policies, reducing the costs associated with frequent strategy changes.

4 Conclusion

This analysis reveals the significant impact of charisma decay coefficients on the choice of management strategies, emphasizing the need to consider the long-term effectiveness and stability of strategies when implementing accident prevention and control management strategies in large public works projects. This provides a crucial theoretical basis for government regulators and project responsible parties when formulating and adjusting management strategies.

This type of analysis helps decision-makers identify and implement more effective management measures to address the complexity and uncertainty in large public works projects. By optimizing the charisma decay coefficient, the overall effectiveness of project management can be improved, thereby achieving a better balance between engineering safety and project benefits.

Date	Low Penalty Mechanism Score	Medium Penalty Mechanism Score	High Penalty Mechanism Score
2023-01-01	1250	1245	1230
2023-01-08	1255	1240	1225
2023-01-15	1260	1245	1235
2023-01-22	1265	1250	1240
2023-01-29	1270	1255	1245
2023-02-05	1275	1260	1250
2023-02-12	1280	1265	1255
2023-02-19	1285	1270	1260
2023-02-26	1290	1275	1265
2023-03-05	1295	1280	1270
2023-03-12	1300	1285	1275
2023-03-19	1305	1290	1280
2023-03-26	1310	1295	1285
2023-04-02	1315	1300	1290
2023-04-09	1320	1305	1295
2023-04-16	1325	1310	1300
2023-04-23	1330	1315	1305
2023-04-30	1335	1320	1310

Scores of Management Strategies Under Low Penalty Mechanism
Under the low penalty mechanism, the scores for management strategies show a continuous upward trend, increasing from 1250 to 1335. This suggests that when the government's supervision of projects is more lenient, project parties tend to adopt autonomous and robust management measures, maintaining stable operations and gradually improving the project. This trend may reflect that, under lower external pressure, project teams can focus more on long-term strategic planning and execution, thereby enhancing the overall effectiveness of project management.

Scores of Management Strategies Under Medium Penalty Mechanism
Under the medium penalty mechanism, scores also show an upward trend but with significant fluctuations (from 1245 to 1320). These fluctuations may indicate that moderate levels of supervision and penalties provide certain pressure and motivation for project parties to adjust and optimize management strategies, but they may also introduce uncertainty and risk in strategy adjustments. This suggests that while moderate penalties can promote management efficiency, excessive instability and frequent changes could negatively impact the project's continuity and predictability.

Scores of Management Strategies Under High Penalty Mechanism

Under the high penalty mechanism, the scores fluctuate most significantly, ranging from 1230 to 1310, reflecting a high level of instability. This may indicate that under strict penalty regimes, project parties frequently change strategies to avoid potential high-cost penalties, leading to a lack of coherence and long-term perspective in management strategies. Such a highly dynamic environment could hinder effective strategy planning and implementation, increasing the complexity and execution costs of project management.

Conclusion and Recommendations

This analysis demonstrates that the effectiveness and stability of management strategies are significantly influenced by the intensity of penalty mechanisms. For large public works projects, choosing the appropriate levels of supervision and penalties is crucial to ensure that projects are executed successfully in a supportive rather than oppressive environment.

Policy Makers should consider setting penalty mechanisms that motivate good behavior without being overly oppressive, thereby encouraging project parties to autonomously implement effective risk management and quality control strategies.

Project Management Teams should maintain adaptability and flexibility in their strategies within a changing regulatory environment while ensuring the coherence and long-term sustainability of these strategies.

Such balanced regulatory strategies not only facilitate the smooth execution of projects but also enhance the morale and motivation of project teams, ultimately leading to a higher success rate of public works projects.

5 Conclusion

In this paper, based on the complex network formed by the interaction behavior of the participating subjects, the adaptive choice of the participating subjects is treated as a complex adaptive system, and an "Agent-cellular Automata" model of the decision-making behavior of the participating subjects in the management of safety accidents in large public works projects is constructed by using the EWA subject learning model, and the simulation of the evolution of safety accident management strategies emergence and competition is carried out. We also simulate the evolution of the emergence and competition of safety accident management strategies. The results of the study show that due to the complexity and uncertainty of the external environment of the project, as well as the limited skills and experience of the participating subjects, the evolution of the safety incident management strategies of the project shows the phenomenon of the mutual influence of multiple safety incident management strategies, and the selection of the safety incident management strategies of each subject is affected by their own expected benefits as well as by the selection of the safety incident management strategies of other heterogeneous subjects in the project. The punishment mechanism of the strategy has a greater impact on the behavior of the participating subjects, and when the punishment mechanism is high, the evolution of the strategy can be oriented towards consistency, so that the project as a whole is in a stable safety state. In addition, the higher the glamour index of the security incident management strategy, the normalization of the strategy will be more obvious. This paper analyses the evolution mechanism of safety incident management strategy selection behavior of heterogeneous subjects in the system, which is of certain guiding significance for the practical promotion and application of safety incident management strategies in large public works. However, there are some shortcomings. This paper did not make an in-depth discussion of other accident prevention and control stakeholders, when there are other project stakeholders involved, the safety accident losses and benefits will be different. Therefore, the follow-up work will be further research on these issues.

References

1. Axelrod, R.: The evolution of cooperation. Basic Books, New York, NY (1984)
2. North, D.C.: Institutions, institutional change, and economic performance. Cambridge University Press, Cambridge, UK (1990)
3. Katz, M.L., Shapiro, C.: Network externalities, competition, and compatibility. Am. Econ. Rev. **75**(3), 424–440 (1985)
4. Camerer, C., Ho, T.-H.: Experience-weighted attraction learning in normal form games. Econometrica **67**(4), 827–874 (1999)
5. Williams, T.: Assessing and moving on from the dominant project management discourse in the light of project overruns. IEEE Trans. Eng. Manage. **52**(4), 497–508 (2005)
6. Sheng, Z.: Complex systems theory for complex project management. Proj. Manag. J. **40**(2), 58–73 (2009)
7. Fehr, E., Gächter, S.: Cooperation and punishment in public goods experiments. Am. Econ. Rev. **90**(4), 980–994 (2000)

O2O Data-Driven Collaborative Optimization of Live Streaming E-commerce for Intangible Cultural Heritage Products

Xiaohu Fan[1,2], Xuejiao Pang[2], Ying Song[1], Jie Han[2], Ruohan Du[2], Mingmin Gong[2], Heying Hu[3], and Beibei Zhang[1(✉)]

[1] Wuhan City Polytechnic, Wuhan 430020, China
{fanxiaohu,songying,zhangbeibei}@whcp.edu.cn
[2] School of Information and Engineering, Wuhan College, Hubei 430070, China
{9452,8201,9093,22202030208}@whxy.edu.cn
[3] Wuchang Shouyi University, Wuhan 430070, Hubei, China

Abstract. Intangible cultural heritage (ICH) products possess distinctive cultural value yet confront four critical bottlenecks in live-streaming e-commerce: prolonged production cycles, elevated logistics and packaging costs, limited market penetration, and the progressive aging of master artisans. Existing literature predominantly addresses isolated design or marketing dimensions and concentrates on standardized goods, leaving a notable research gap in holistic, data-driven frameworks that simultaneously optimize the entire "production–sales–logistics" chain under live-streaming contexts. Moreover, prior work has not adequately resolved the challenges of data sparsity and pronounced seasonal demand volatility inherent to non-standard ICH offerings.

To bridge these gaps, this study proposes a threefold contribution: (1) a data-driven collaborative optimization framework that integrates online live-streaming platforms with offline experiential touchpoints; (2) a hybrid demand-forecasting pipeline combining CNN-LSTM for short-term demand fluctuations and Temporal Fusion Transformers (TFT) for long-term trend prediction; and (3) a hybrid recommendation algorithm that fuses collaborative filtering with content-based techniques to mitigate data sparsity while preserving cultural relevance.

Empirical validation demonstrates that the proposed framework significantly outperforms traditional baselines in forecasting accuracy (RMSE and MAE) and elevates click-through and purchase conversion rates, thereby enhancing both commercial viability and cultural preservation.

Keywords: Intangible Cultural Heritage E-commerce · O2O Supply Chain Optimization · Temporal Fusion Transformer · CNN-LSTM · Personalized Recommendation Algorithm

1 Introduction

Intangible cultural heritage is an important treasure of human civilization [1]. With the acceleration of globalization and the transformation of modern lifestyles, many intangible cultural heritage projects are facing difficulties in inheritance, low market awareness, and poor economic benefits. In this context, e-commerce, as an emerging business model, provides new opportunities for the promotion and sales of intangible cultural heritage products [2]. Especially in the O2O (Online to Offline) model, through the integration of online and offline, benefits from combine expand the market influence of intangible cultural heritage products and provide consumers with a better shopping experience [3].

1.1 Current Situation and Challenges

ICH products have unique cultural connotations and artistic value, but due to their complex production processes and long cycles, their supply efficiency is relatively lower than industry pipelines by 90% plus. In the process of market expansion, intangible cultural heritage products face four major challenges [4]:

Long production cycle and easy out of stock promotion: The production of ICH products often requires multiple traditional craftsmanship processes [5], which are time-consuming. During e-commerce promotional activities, the order volume may surge in a short period of time, and due to production cycle limitations, it is difficult to replenish in a timely manner, resulting in stockouts and affecting sales conversion rates and consumer satisfaction.

High shipping packaging costs and high return rates: ICH products come in various forms, some of which have special materials and exquisite decorations, requiring customized packaging materials and fine techniques [6], which undoubtedly increases packaging costs. Meanwhile, due to the potential discrepancy between consumers' expectations for ICH products and the actual products received, the return rate is relatively higher about 30%, further increasing operational costs.

Limited market awareness and small consumer group: The cultural connotation and artistic value of ICH products need to be deeply disseminated and promoted [7].

Aging of inheritors and insufficient innovation drive: Most inheritors of ICH are elderly, and the younger generation has lower interest and participation in intangible cultural heritage [8]. The inheritor team is facing a crisis of discontinuity, insufficient innovation drive, and difficulty in meeting market demand.

1.2 Strength of O2O Mode

The O2O model provides new practical ideas for promoting ICH products and enhancing consumer experience through the integration of online and offline channels. Utilize live streaming and social media online to attract users [9], which optimize product selection strategies by analyzing user behavior data. Offline, holographic cabin technology is used to display product details, enhancing consumers' cultural experience and purchase intention.

1.3 Motivation

The aim of this study is to construct a collaborative optimization framework for the supply chain of live streaming e-commerce of intangible cultural heritage products under the O2O model, to achieve information sharing and collaborative operation at all stages, improve customer satisfaction and market competitiveness, and achieve a win-win situation of cultural value and economic benefits.

This practical significance mainly reflected in the following three aspects: firstly, providing new ideas and methods for the market promotion and ICH products, and promoting their sustainable development in modern society; Secondly, by leveraging big data and precision marketing technologies, we can enhance the market adaptability and competitiveness of ICH products, and help to realize their commercial value; The third is to explore the application of O2O model in the e-commerce of ICH products, providing useful references and inspirations for optimizing the e-commerce supply chain of other similar products, and promoting the entire industry.

2 Related Works

Domestic and foreign scholars' research on the e-commerce of intangible cultural heritage products focuses on digital transformation, cultural value mining, and supply chain optimization [10]. The annual transaction volume of intangible cultural heritage e-commerce has exceeded 100 billion yuan, and the proportion of young consumers has significantly increased. However, there are prominent problems such as high return rates (up to 15%–20%) and slow supply chain response speed.

AIGC technology (such as generative AI) is used for design innovation of non-legacy cultural and creative products [11], but the production end still relies on manual production, resulting in capacity fluctuations and inventory imbalances.

The non-standard attributes of intangible cultural heritage products, such as long handmade production cycles and high packaging costs, make it difficult for traditional e-commerce models to adapt, and require the integration of online and offline resources through O2O models.

Japanese scholars have proposed a "traditional craftsmanship plus subscription based e-commerce" model to alleviate inventory pressure through a pre order mechanism, but it has not solved the problem of personalized recommendations [12]. Existing achievements mostly focus on a single link (such as design or marketing), lacking systematic research on the collaborative optimization of the entire "production sales logistics" chain in live streaming e-commerce scenarios.

2.1 O2O Mode in E-commerce Supply Chain

The O2O model collects user behavior data through offline experiences (such as holographic cabin displays), which feeds back into online product selection and inventory forecasting. For example, Taotian Group achieved a 30% increase in user conversion rate through "online live streaming plus offline pop-up stores". B2B e-commerce of cultural tourism, big data and AI technology have been applied to dynamic pricing and path optimization, but the field of intangible cultural heritage is still in the exploratory stage [13].

2.2 Algorithms and Models

Collaborative filtering algorithm: Used for cross channel user interest mining, but the long tail feature of intangible cultural heritage products leads to data sparsity issues [13]. The spatiotemporal prediction model, such as the Temporal Fusion Transformer (TFT), performs well in predicting the demand for fast-moving consumer goods, but needs to adapt to the seasonal fluctuations of intangible cultural heritage products (such as the surge in sales during traditional festivals). Existing O2O research mostly focuses on standardized products and lacks algorithm design for adapting non-standard attributes of intangible cultural heritage products [14].

This article intends to fill the research gap in the field of intangible cultural heritage live streaming e-commerce supply chain by constructing a three in one optimization framework of "data-driven O2O collaboration cultural embedding", providing a new paradigm for balancing cultural inheritance and commercial value [15].

3 Architecture and Data

3.1 Flowchart and Architecture

The architecture of the O2O e-commerce supply chain collaborative optimization system for intangible cultural heritage products constructed in this study mainly includes the following core modules as shown in Fig. 1:

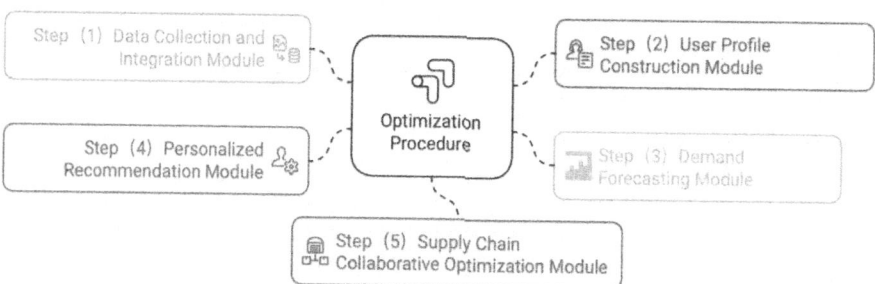

Fig. 1. Workflow of Optimization Procedure

Data collection and integration module: responsible for collecting user behavior data, sales data, product information, etc. from multiple channels such as online live streaming platforms, offline holographic cabin display systems, and e-commerce platforms, and conducting data cleaning and preliminary integration.

User Profile Construction Module: Based on collected multi-source data, precise user profiles are constructed using techniques such as feature engineering and clustering analysis, providing a foundation for personalized recommendations and precision marketing.

Demand forecasting module: Combining the advantages of CNN-LSTM model and TFT model, predict the short-term and long-term demand for intangible cultural heritage products. Among them, the CNN-LSTM model is mainly used to process real-time online live streaming sales data and capture short-term demand fluctuations; The TFT model focuses on analyzing complex offline holographic cabin display data and historical sales data, providing long-term demand trend predictions.

Personalized recommendation module: Based on user profiles and demand prediction results, collaborative filtering algorithms are used to provide personalized recommendations for intangible cultural heritage products to improve user purchase conversion rates.

Supply chain collaborative optimization module: Feedback demand forecasting results to inventory management and logistics distribution systems, collaborate to optimize online and offline inventory levels and logistics routes, reduce operating costs, and improve supply chain efficiency.

3.2 Data Sources and Acquisition Methods

Online Data. The online data mainly comes from live streaming platforms for intangible cultural heritage products and e-commerce platforms. By collaborating with the platform, obtain users' basic information, behavioral data, and transaction records. Specifically, it includes:

User basic information: Obtain username, age, gender, contact information, etc. through the user registration interface.

User behavior data: Use behavior tracking interfaces to collect users' viewing history, interactive behavior (such as likes, comments, shares), and purchase records on the platform.

Product data: Extract detailed information about intangible cultural heritage products from the product database, including product types, styles, price ranges, etc.

Offline Data. Offline data mainly comes from physical stores and holographic displays of intangible cultural heritage products. Collect users' purchase records, preference feedback, and participation in activities through the membership system and POS system. At the same time, use in store cameras and sensors to collect information on users' behavior trajectories and stay time in the store. Details of user, including basic information, behavior, preference and social relations shown in Tables 1, 2, 3 and 4 below.

Table 1. User Basic Information

Field Name	Data Type	Description
user_id	INT	User Unique Identifier
username	VARCHAR(50)	Username
age	INT	Age
gender	CHAR(1)	Gender (M/F)
phone	VARCHAR(20)	Mobile Phone Number
email	VARCHAR(100)	Email Address
address	VARCHAR(200)	Address

Table 2. User Behavior

Field Name	Data Type	Description
behavior_id	INT	Unique Identifier of Behavior Record
user_id	INT	User Unique Identifier
behavior_type	VARCHAR(50)	Behavior Type
behavior_time	DATETIME	Time of Behavior Occurrence
behavior_detail	VARCHAR(200)	Behavior Details
watch_duration	INT	Duration of Watching (in seconds)
purchase_amount	DECIMAL(10,2)	Purchase Amount

Table 3. User Preference

Field Name	Data Type	Description
preference_id	INT	Unique Identifier of Preference Record
user_id	INT	User Unique Identifier
product_type	VARCHAR(50)	Favorite Types of Intangible Cultural Heritage Products
product_style	VARCHAR(50)	Favorite Styles of Intangible Cultural Heritage Products
preferred_material	VARCHAR(50)	Preferred Materials
preferred_color	VARCHAR(50)	Preferred Colors
price_range	VARCHAR(50)	Preferred Price Range
favorite_designer	VARCHAR(100)	Favorite Inheritors of Intangible Cultural Heritage or Designers

Table 4. User Social Relation

Field Name	Data Type	Description
relation_id	INT	Unique Identifier of Relationship Record
user_id	INT	User Unique Identifier
friend_id	INT	Unique Identifier of Friend
relation_type	VARCHAR(50)	Relationship Type (friend, follow, fan, etc.)
interaction_frequency	INT	Interaction Frequency (Interactions per Month)
last_interaction_time	DATETIME	Last Interaction Time

3.3 Data Acquisition and Interface Design

This study has established a comprehensive data collection and management system for O2O e-commerce of intangible cultural heritage products. Obtain user basic information, viewing history, interactive behavior, purchase records, and social relationships through live streaming platforms, e-commerce platforms, and social media APIs online; Offline collection of member information, in store purchase records, and user behavior trajectories through membership systems, POS systems, and sensor data interfaces. The collected data undergoes a cleaning step to remove outliers and missing values, ensuring completeness and accuracy, and accelerating model convergence through normalization processing. At the same time, this study attaches great importance to data security and privacy protection, adopting encryption technology and access control mechanisms to ensure data security and user privacy.

4 Methods

The construction of user profiles is the foundation for achieving precision marketing and personalized recommendations. By deeply mining multi-source data, static features (such as age, gender, region, etc.) and dynamic behavioral features (such as live streaming duration, interaction frequency, purchase history, etc.) of users are extracted, and combined with the cultural attributes and consumption characteristics of intangible cultural heritage products, clustering analysis and other techniques are used to divide users into different groups. Each group has similar consumption behaviors and preferences, providing a basis for subsequent personalized recommendations and marketing strategies. Flowchart of the chapter is shown in Fig. 2.

4.1 Implementation of Personalized Recommendation Algorithm

Based on the constructed user profile and demand prediction results, personalized recommendation algorithms provide users with highly matched intangible cultural heritage product recommendations that match their interests and needs through collaborative filtering and other technologies. Specifically, for each user group, products that meet their preferences are selected from the product library based on their characteristics and behavior patterns, and combined with inventory and logistics capabilities in demand forecasting, the final recommendation list is generated. At the same time, in order to improve the diversity and novelty of recommendations, content-based recommendation methods are introduced, taking into account factors such as cultural connotations and technological characteristics of products, to provide users with richer choices.

4.2 CNN-LSTM Model

The CNN-LSTM model combines the local feature extraction ability of CNN with the modeling ability of LSTM for time series. The specific principle is as follows:

CNN layer: uses one-dimensional convolution operation to extract local features of time series data. Convolutional kernels slide over time steps to generate feature maps and capture short-term dependencies in the data.

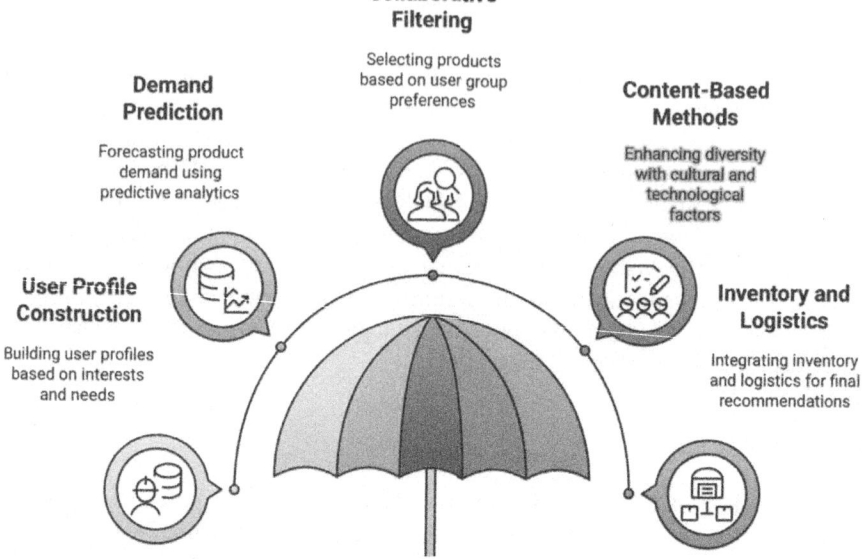

Fig. 2. Components of ICH products Recommendation Algorithm

The formula is expressed as:

$$C = Conv1D(X, filters, kernel\_size, activation) \quad (1)$$

Among them, X is the input data, filters is the number of convolution kernels, kernel_2 is the size of convolution kernels, activation is the activation function, and C is the feature map after convolution.

LSTM layer: takes the feature sequence extracted by CNN as input, captures long-term dependencies through gating mechanism, and outputs a hidden state sequence.

$$H_t = LSTM(C_t, H_{t-1}) \quad (2)$$

Among them, C_t is the feature input of time step t, H_{t-1} is the hidden state of the previous time, and H_t is the hidden state of the current time.

Fully connected layer: Maps the output of LSTM to the predicted value space through a fully connected layer. The formula is expressed as:

$$Y_{pred} = Dense(H_T) \quad (3)$$

Among them, H_T is the hidden state of LSTM at the last moment, and Y_{pred} is the final predicted value.

Before applying the CNN-LSTM model, preprocess the sales data of intangible cultural heritage products:

Data cleaning: Remove outliers and missing values to ensure the integrity and accuracy of the data.

Time series construction: Convert raw sales data into a supervised learning problem. Set the time window and prediction lag, and construct input-output sample pairs. The formula is expressed as:

$$\text{Input} = \{X_t | t = 1, 2, ..., N - \text{lag}\} \tag{4}$$

$$\text{Output} = \{Y_t | t = 1 + \text{lag}, 2 + \text{lag}, ..., N\} \tag{5}$$

Among them, X_t is the input feature of time step t, Y_t is the corresponding target value, and N is the total number of time steps.

Data normalization: Normalize the data to distribute it in the [0,1] interval, accelerating model convergence.

4.3 Model Construction and Training

Model Construction

CNN layer: Set the number of convolution kernels to 64, kernel size to 1, and activation function to ReLU.

LSTM layer: The number of hidden units is 50, and the activation function is ReLU.

Fully connected layer: The output layer uses a linear activation function to predict sales volume for the next 90 days. The model structure is as follows:

$$X \xrightarrow{Conv1D} MaxPooling1D \xrightarrow{Flatten} LSTM \xrightarrow{Dense} Y_{pred} \tag{6}$$

Model Training

Loss function: Mean Square Error (MSE) is used as the loss function £, which with results would be discussed in later chapters.

$$\mathcal{L} = \frac{1}{N} \sum_{i=1}^{N} (Y_{\text{pred},i} - Y_{\text{true},i})^2 \tag{7}$$

Optimizer: Use Adam optimizer with a learning rate of 0.001.

Training process: Divide the dataset into a training set and a validation set, with a ratio of 6:4. Train for 50 epochs with a batch size of 32.

4.4 Temporary Fusion Transformer (TFT) Model

Accurate demand forecasting is crucial for optimizing inventory and logistics in the e-commerce supply chain management of intangible cultural heritage products. This article introduces the TFT model and combines the characteristics of intangible cultural heritage product sales data to achieve accurate prediction of demand fluctuations.

Variable Selection Network. The input layer of TFT processes two types of inputs: time series input and static input. Time series inputs include historical sales data, promotional activities, etc., while static inputs include product types, price ranges, etc. The embedding layer maps classification features (such as product types) to dense vector representations so that the model can better handle these features. Variable selection network is one of the core components of TFT, implemented based on Gated Residual Network (GRN). For each time step t and each feature f, calculate the importance weight α of the feature

$$\alpha_{t,f} = \sigma\left(W_g \cdot tanh\left(W_f \cdot X_{t,f} + b_f\right) + b_g\right) \tag{8}$$

X is the input of feature f at time step t, W is a learnable parameter, and σ is the sigmoid function. This weight determines the contribution of feature f to the prediction at time step t.

Self Attention Mechanism. The self attention mechanism allows the model to focus on the relationships between different time steps. For time steps t and t', Calculate the attention score e_t, t':

$$e_{t,t'} = Q_t \cdot K_{t'}^T \tag{9}$$

Q_t and $K_{t'}$ are the query vector of time step t' and the key vector of time step t' respectively. The attention score is normalized by the $soft_{max}$ function to obtain the attention weight a_t, t', Used for weighted summation to generate context vector V_t:

$$\alpha_{t,t'} = \frac{esp(e_{t,t''})}{\sum_{t'} esp(e_{t,t''})} \tag{10}$$

$$V_t = \sum_{t'} \alpha_{t,t'} \cdot V_{t'} \tag{11}$$

The gated residual network combines the residual connection and gating mechanism to control the flow of information, which is basically the same as the previous LSTM part. The output layer maps the output of the decoder to the predicted value.

$$\widehat{y}_{t+\tau} = W_{out} \cdot H_{t+\tau} + b_{out} \tag{12}$$

For the future time step $t + \tau$, the predicted value $\widehat{y_{t+\tau}}$ is generated through the full connection layer, where W_{out} and b_{out} is the weight and bias parameter of the output layer.

4.5 Temporary Fusion Transformer (TFT) Model

Personalized recommendation algorithms aim to provide customized recommendations for intangible cultural heritage products based on users' interests and behavior patterns. This study adopts collaborative filtering algorithm as the core recommendation strategy and combines it with content-based methods for optimization. Collaborative filtering algorithms analyze user behavior data, such as purchase history, browsing history, interactive behavior, etc., to uncover implicit associations between users and products. Specifically, for the target user, the algorithm first obtains its feature vector in the user profile construction module and then searches for neighboring users with similar features in the user product interaction matrix, generating a recommendation list based on the preferences of neighboring users. This article mainly adopts online data for prediction, integrates offline data feedback, and accurately predicts demand and user profiles.

5 Result and Discussion

The root mean square error (RMSE) and mean absolute error (MAE) were used as the evaluation indexes of the model performance. These indicators can effectively measure the difference between the predicted value and the real value. RMSE is more sensitive to large errors, while Mae treats all errors equally. The combination of the two can comprehensively evaluate the prediction accuracy of the model.

$$\text{RMSE} = \sqrt{\frac{1}{N}\sum_{i=1}^{N}|(Y_{\text{pred},i} - Y_{\text{true},i})^2} \tag{13}$$

$$\text{MAE} = \frac{1}{N}\sum_{i=1}^{N}|Y_{\text{pred},i} - Y_{\text{true},i}| \tag{14}$$

5.1 Current Situation and Challenges

Training and verification loss: through the visual training and verification loss curve, it is observed that the loss of CNN-LSTM model gradually decreases during the training process and tends to be stable after about 30 epochs, indicating that the model has basically converged. The RMSE and Mae of the training set were 12.34 and 8.65, respectively; The RMSE on the validation set is 14.78 and the MAE is 10.23. This shows that the model has good prediction performance in both the training set and the validation set, and has strong generalization ability.

Analysis of forecast results: further analysis of the forecast results shows that CNN-LSTM model has high accuracy in dealing with short-term demand fluctuations, and can better capture the local characteristics and short-term trends in sales data. For example, during the promotion activities, the model can accurately predict the peak sales, which provides strong support for inventory preparation and logistics scheduling.

5.2 TFT Model Results

Training and verification loss: the training and verification loss curve of TFT model showed a relatively smooth downward trend, and reached a stable state after about 40 epochs. The RMSE and Mae of the training set were 9.87 and 6.54, respectively; The RMSE on the validation set is 11.32 and the MAE is 7.89. This shows that TFT model has better fitting ability and generalization performance when dealing with complex multivariate time series data. The procedure of TFT model shown in Fig. 3.

Analysis of forecast results: TFT model shows significant advantages in long-term demand forecasting, and can effectively capture long-term dependencies and global patterns in data. For the sales trend of intangible cultural heritage products in different seasons and different promotional activities, TFT model can provide more accurate prediction results, and provide a reliable basis for production planning and inventory management.

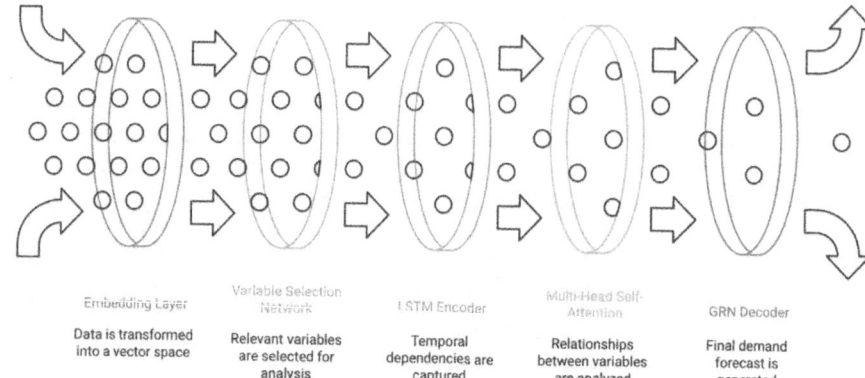

Fig. 3. Temporal Fusion Transformer Process

Both metrics (RMSE and MAE) are computed on the normalized sales volume of intangible cultural heritage products. Results shown in Table 5 below. The CNN-LSTM network exhibits faster convergence and excels at capturing localized, high-frequency patterns, whereas the TFT architecture demonstrates superior generalization for extended temporal horizons and complex multivariate dependencies.

Table 5. Comparative Performance of CNN-LSTM and TFT on Forecasting

Model	Training RMSE	Training MAE	Validation RMSE	Validation MAE	Convergence Epoch
CNN-LSTM	12.34	8.65	14.78	10.23	~30 epochs
TFT	9.87	6.54	11.32	7.89	~40 epochs

Table 6. Performance of the Personalized Recommendation Algorithm

Metric	Hybrid Algorithm	No-Recommender Baseline	Relative Δ
Click-Through Rate (CTR)	18.5%	12.8%	+45%
Purchase Conversion Rate (CVR)	6.3%	4.8%	+32%
User Satisfaction (CSAT ≥ 4/5)	70.8%	58.1%	+12.7 pp

Notes. CTR and CVR are reported on the hold-out test set; CSAT is derived from a post-interaction survey (response rate = 23.6%). Statistically significant improvements are confirmed by χ^2 tests ($p < 0.01$).

5.3 Personalized Recommendation Algorithm Results

Recommendation accuracy: through offline evaluation and online a/b test, personalized recommendation algorithm performs well in recommendation accuracy. In the test set, the click through rate of recommended products was 18.5%, and the purchase conversion rate was 6.3%, which increased by 45% and 32% respectively compared with the case without the recommendation algorithm.

User satisfaction: the user feedback survey shows that more than 70% of users are satisfied with the personalized recommendation results and believe that the recommended products meet their interests and needs. This shows that personalized recommendation algorithm not only improves the sales conversion rate, but also enhances the shopping experience and satisfaction of users (Table 6).

5.4 Discussion

The comprehensive experimental results show that cnn-lstm model and TFT model have their own advantages in the demand prediction of intangible cultural heritage products. Cnn-lstm model performs well in short-term demand forecasting, can quickly respond to market changes, and is suitable for online live broadcasting with high real-time requirements; The TFT model has more advantages in long-term demand forecasting, and can provide accurate forecasting support for offline production and inventory management. By combining collaborative filtering and content-based methods, personalized recommendation algorithm effectively improves the accuracy and diversity of recommendation, meets the personalized needs of users for intangible cultural heritage products, and promotes the marketing and cultural heritage of intangible cultural heritage products.

However, during the experiment, we also found some areas that need further improvement. For example, in the case of high data sparsity, the performance of personalized recommendation algorithm may be affected; For some emerging intangible cultural heritage products, the accuracy of the demand forecasting model needs to be improved. Future research will focus on optimizing the algorithm, combining more external data and context information to further improve the prediction ability and recommendation effect of the model.

6 Result and Discussion

Aiming at the collaborative optimization of o2o e-commerce supply chain for intangible cultural heritage products, this study constructed a system architecture integrating a variety of advanced technologies and algorithms. Through the construction of accurate user portraits, in-depth mining of user characteristics and preferences; The cnn-lstm model and TFT model are used to predict the short-term and long-term demand respectively, and the personalized recommendation algorithm is combined to realize the precise marketing and supply chain optimization of intangible cultural heritage products. The experimental results show that the proposed method performs well in demand forecasting accuracy and personalized recommendation effect, and effectively improves the market competitiveness and cultural heritage of intangible cultural heritage products.

Multi source data fusion: integrate multi-channel data such as online live broadcast and offline holographic cabin display to provide comprehensive data support for user portrait and demand prediction.

Hybrid model application: creatively combine cnn-lstm and TFT models, give full play to their advantages in short-term and long-term demand forecasting, and improve the forecasting accuracy.

Personalized recommendation optimization: a personalized recommendation algorithm combining collaborative filtering and content-based methods, which not only considers user behavior, but also integrates the cultural attributes of intangible cultural heritage products to improve the recommendation effect.

Future research directions for this study include enhancing data quality through refined preprocessing to mitigate noise effects on model performance. Additionally, optimizing the CNN-LSTM and TFT model architectures by incorporating external factors such as market trends and consumer behavior could further improve prediction accuracy and practical applicability. The exploration of hybrid algorithmic frameworks that combine advanced deep learning techniques with traditional statistical methods is also crucial for addressing complex market demands. Finally, integrating the developed system architecture with real-world O2O e-commerce platforms for intangible cultural heritage products and conducting large-scale practical applications will be vital for advancing the intelligent transformation of this industry.

Acknowledgements. This research has received substantial support from Digital Twin Smart Commerce Application Research Center of WHCP and the 2025 Hubei Provincial Chinese Vocational Education Society Research Project, "Holographic Technology and Collaborative Filtering Algorithm Empowering Digital Protection of Hubei Intangible Cultural Heritage—Innovations in User Profiling and Online Store Operations Teaching."

Funding. This project was supported by the 2025–2026 China Association for Adult Education Intangible Cultural Heritage Research Project, "Holographic Cabin Empowering Digital Preservation of Yingshan Entwined Flower and O2O Immersive Interaction Mode Research"; the 2019 Hubei Provincial Department of Education Philosophy and Social Sciences Research Project, "Research on the Information-based Inheritance of Jingchu Culture" (No. 19G095).

Conflicts of Interest. The authors declare no conflicts of interest.

References

1. Lin, H.: Evaluation on the protection and development of intangible cultural heritage in she township, Jingning from the perspective of ecological civilization. Sustainability **15**(3), 2330 (2023)
2. Yingqing, X., Hasan, N.A.M., Jalis, F.M.M.: Purchase intentions for cultural heritage products in E-commerce live streaming: an ABC attitude theory analysis. Heliyon **10**(5) (2024)
3. Halkos, G.E., Koundouri, P.C., Aslanidis, P.S.C., et al.: Evaluating the tangible and intangible parameters of cultural heritage: an economic meta-analysis in a global context. Discover Sustain. **5**(1), 187 (2024)

4. Giglitto, D., Ciolfi, L., Bosswick, W.: Building a bridge: opportunities and challenges for intangible cultural heritage at the intersection of institutions, civic society, and migrant communities. Int. J. Herit. Stud. **28**(1), 74–91 (2022)
5. Li, X.Z., Chen, C.C., Kang, X.: The design and evaluation of teaching activities combining traditional lantern craftsmanship with primary education in the perspective of intangible cultural heritage. In: 2021 3rd International Conference on Computer Science and Technologies in Education (CSTE), pp. 53–57. IEEE (2021)
6. Zu-zhou, Q., Hui, Q., Yun, L.: Packaging design of intangible cultural heritage food based on the characteristics of traditional folk houses in east Fujian. Food Mach. **38**(7), 138–143 (2022)
7. Tan, N., Anwar, S., Jiang, W.: Intangible cultural heritage listing and tourism growth in China. J. Tour. Cult. Chang. **21**(2), 188–206 (2023)
8. Guan, Z.: Digital rescue protection of representative inheritors of intangible cultural heritage in the information age. J. Phys.: Conf. Ser.. IOP Publishing **1744**(4), 042124 (2021)
9. Hou, Y., Kenderdine, S., Picca, D., et al.: Digitizing intangible cultural heritage embodied: State of the art. J. Comput. Cult. Heritage **15**(3), 1–20 (2022)
10. Tavares, D.S., Alves, F.B., Vásquez, I.B.: The relationship between intangible cultural heritage and urban resilience: a systematic literature review. Sustainability **13**(22), 12921 (2021)
11. Wang, W., Yang, G., Liu, X.: Reflections on AIGC empowering the development of guizhou's intangible cultural heritage. J. Soc. Sci. Hum. Literat. **7**(3), 38–41 (2024)
12. Qiu, Q., Zuo, Y., Zhang, M.: Intangible cultural heritage in tourism: Research review and investigation of future agenda. Land **11**(1), 139 (2022)
13. Swoboda, B., Müller, M.: E-commerce firms' geographic scope: roles of intangible resources and country-specific moderators. Market. ZFP-J. Res. Manage. **44**(1) (2022)
14. Ranjan, A., Chaturvedi, .P.: Digital technologies and the intangible cultural heritage of the rural destination. In: Disruptive Innovation and Emerging Technologies for Business Excellence in the Service Sector, pp. 196–218. IGI Global Scientific Publishing (2022)
15. Alahmari, N., Mehmood, R., Alzahrani, A., et al.: Autonomous and sustainable service economies: data-driven optimization of design and operations through discovery of multi-perspective parameters. Sustainability **15**(22), 16003 (2023)

Deep Learning-Based Evaluation of High-Quality Economic Development in China

Xuefen Chen[1,2], Simeng Su[1,2(✉)], and Yue Zhang[1,2]

[1] School of Computer and Information Science, QingHai Institute of Technology, Xining 810016, China
337905565@qq.com, zhangyue@jju.edu.cn
[2] School of Computer and Big Data Science, Jiujiang University, Jiujiang 332005, China

Abstract. This paper constructs an evaluation model of China's high-quality economic development based on deep learning methods, and applies kernel density estimation and Markov chains to clarify the spatiotemporal dynamic evolution and spatial development trends of high-quality economic growth. The study reveals the following findings: First, it can be observed from the changes in the color of the heat map that, over time, the high-quality economic development index of most cities shows an upward trend; however, differences persist in the growth rates and volatility among cities. Second, a comparative analysis between modern and traditional methods (entropy weight method and equal weighting method) indicates that the high-quality economic development index calculated using deep learning outperforms traditional linear evaluation approaches. Third, based on the kernel density estimation graphs (which can reveal the trends of the dynamic evolution of high-quality economic development in various regions), the research results show that the levels of high-quality economic development in China's four major regions (eastern, central, western and northeastern) have improved to varying degrees over time. Fourth, the spatial evolution trend of China's high-quality economic development, revealed through Markov chain analysis, shows that China's high-quality economic development is significantly influenced by spatial spillover effects from neighboring areas. However, the direction and intensity of this influence depend on the economic development gaps between regions. When these gaps are small, there is a significant mutual promotion effect; but when the gaps become too large, the driving effect of high-level regions on low-level regions is constrained.

Keywords: High-quality economic development · Deep learning · Kernel density estimation · Markov chain

1 Introduction

With the rapid development of artificial intelligence (AI), deep learning, as a core branch of AI, has demonstrated outstanding performance and broad application prospects in various fields. Especially in areas such as risk identification, sentiment classification, and land price evaluation, deep learning has provided new perspectives and methodologies for addressing complex socio-economic issues through its powerful data processing and pattern recognition capabilities [1,2].

Since the reform and opening-up policy was implemented, China's economic growth in terms of quantitative indicators has drawn global attention. In 2024, China's GDP reached 18.94 trillion U.S. dollars, an increase of 87 times compared to 21.65 billion dollars in 1978. However, economic growth is stage-specific, and the stage at which economic growth currently stands constitutes an essential aspect of a country's national conditions. From the early focus on "economic growth rate as the sole criterion for promotion competition", to the subsequent emphasis on "improving the quality and efficiency of economic growth", and now to the current era's focus on "high-quality development", China has undergone significant transitions in its development philosophy.

Since the concept of high-quality development was introduced, it has attracted sustained attention from both domestic and international scholars. The research findings mainly fall into the following two categories: First, qualitative studies. These focus on the connotation and characteristics of high-quality development [3], its internal mechanisms [4], and pathways to realization [5]. Second, the measurement and evaluation of the level of high-quality economic development. Some studies use single indicators to reflect certain aspects of development, while others conduct comprehensive measurements by constructing evaluation indicator systems for high-quality economic development. On one hand, some studies employ single indicators to reflect the level of high-quality development from specific perspectives, such as total factor productivity [6], ecological welfare performance, or green development welfare [7,8]. However, these single variables are insufficient to fully capture the rich connotations and multidimensional characteristics of high-quality development. At the same time, factors such as resource utilization efficiency [9], economic volatility [10], and regional income inequality [11] are also commonly used as evaluation criteria. On the other hand, in recent years, an increasing number of studies have attempted to construct comprehensive indicator systems to evaluate high-quality development, aiming to overcome the limitations of single-indicator approaches. This type of research mainly includes two aspects: one is the derivative analysis based on the measurement of economic growth quality, and the other is the comprehensive evaluation grounded in the new development philosophy. The first type extends from the evaluation index system of economic growth quality, incorporating multiple factors such as environment, society, and economy [12,13]. The second type constructs an evaluation system based on the five development concepts of innovation, coordination, environmental friendliness, openness, and shared prosperity [14,15]. However, most of the above studies rely heavily on traditional economic evaluation methods such as Principal Component Analysis (PCA) and Entropy-weighted Technique for Order Preference by Similarity to Ideal Solution (Entropy-TOPSIS). These methods often struggle when dealing with large volumes of complex data. Therefore, this study aims to apply deep learning techniques to comprehensively evaluate the level of high-quality economic development.

Compared with existing literature, the main contributions of this paper are as follows: First, it applies deep learning methods for the first time to construct an evaluation model of high-quality economic development, overcoming the subjectivity and arbitrariness in weight assignment inherent in traditional label-free data measurement approaches. Second, by introducing deep learning techniques, it further expands the "toolkit" for quantitative research on high-quality economic development, providing new insights and methodologies for research in this field. Third, by employing kernel density estimation and Markov chain analysis, the study clarifies the spatiotemporal dynamic evolution and spatial development trends of high-quality economic development. This allows for predictions of future spatial patterns and provides scientific support for policymakers to formulate effective measures that promote regional balanced development.

2 Research Methodology

2.1 Construction of the Evaluation Index System for High-Quality Economic Development

The Outline of the 14th Five-Year Plan clearly states that high-quality development is the kind of development that meets the people's growing needs for a better life. It represents development that embodies the new development philosophy-innovation as the primary driving force, coordination as an inherent feature, green development as a common form, openness as an inevitable path, and shared prosperity as the ultimate goal. The new development philosophy centered on innovation, coordination, green, openness, and shared benefits was proposed by the Party Central Committee with Comrade Xi Jinping at its core. It is based on a profound summary of domestic and international development experiences and lessons, and a thorough analysis of global development trends. Therefore, constructing a scientific and reasonable indicator system is a necessary prerequisite for accurately evaluating the level of high-quality economic development in Chinese cities.

Building upon an in-depth analysis of China's national development strategies and planning content outlined in the 14th Five-Year Plan, this study adopts the indicator construction framework proposed by [16], to construct a comprehensive evaluation index system for urban high-quality development in China. This system includes five dimensions: innovative development, coordinated development, green development, open development, and shared development, as shown in Table 1.

Innovative Development: Innovation is the primary driving force for development and a strategic pillar for building a modernized economic system. A modernized economic system is a key feature of the high-quality development stage, and innovation, as its core support, aims to uphold a higher-quality economy.

Table 1. Evaluation Index System for High-Quality Economic Development in China. Note: In the "Attribute" column, "+(-)" indicates that the measurement index is a positive (negative) indicator under the specified evaluation method, meaning that a higher (lower) value is preferable.

Target	Subsystem (Weight)	Criteria Layer	Specific Measurement Indicators	Indicator Measurement Method	Attribute	Code
High Quality Economic Development	Innovation	Knowledge Funding Efficiency	Science and Education Funding Effect	Each City's GDP/Science and Education Expenditure	+	C1
		Information Industry Personnel Input Intensity	Information Service Personnel Input Intensity	Per Capita Information Transmission Software and IT Service Personnel /Total Population	+	C2
		Knowledge Personnel Input Intensity	Input Intensity of Students in Regular Higher Education Institutions	Number of Students in Regular Higher Education Institutions / Total Number of Students	+	C3
	Coordination	Regional Coordination	Regional Income Coordination	Per Capita GDP of Each City/ Per Capita GDP of the Province	+	C4
		Industrial Coordination	Industrial Structure Coordination	Output Value of the Tertiary Industry/ the Secondary Industry	+	C5
		Financial Coordination	Financial Deposit and Loan Coordination	Balance of Financial Deposits / Balance of Financial Loans	+	C6
	Green Development	Three Wastes Emission	Waste Water Discharge	Industrial Waste Water Discharge /GDP of Each City	+	C7
			Waste Gas Emission	Industrial Waste Gas Emission/GDP of Each City	+	C8
			Smoke and Dust Emission	Industrial Smoke and Dust Emission/GDP of Each City	+	C9
		Pollutant Treatment	Comprehensive Utilization Rate of General Industrial Solid Waste	Direct Data Acquisition	+	C10
			Centralized Treatment Rate of Sewage Treatment Plants	Direct Data Acquisition	+	C11
			Harmless Treatment Rate of Domestic Garbage	Direct Data Acquisition	+	C12
		Green Environment	Per Capita Park Green Area Per 10,000 People	Direct Data Acquisition	+	C13
			Green Coverage Rate of Built-up Area	Direct Data Acquisition	+	C14
	Openness	Foreign Investment Openness	Actual Utilization of Foreign Investment	Actual Foreign Investment / GDP of Each City	+	C15
		Foreign Trade Openness	Goods Import and Export Volume	Total Import and Export Trade / GDP of Each City	+	C16
		Information Openness	Number of International Internet Users	Direct Data Acquisition	+	C17
	Sharing	Completeness of Education Facilities	Per Capita Public Education Funding	Education Expenditure of Each City / Total Population	+	C18
		Completeness of Cultural Facilities	Per Capita Collection of Public Libraries	Direct Data Acquisition	+	C19
		Completeness of Medical Facilities	Per Capita Number of Beds in Medical and Health Institutions	Number of Beds in Medical and Health Institutions / Total Population	+	C20

Innovation emphasizes driving economic growth through various forms of innovation, including technological, managerial, and business model innovation. Only by continuously promoting innovation can we improve production efficiency, enhance international competitiveness, and achieve optimization and upgrading of the economic structure. This study measures urban innovative development from three aspects: efficiency of knowledge-related funding, intensity of information industry personnel input, and intensity of knowledge personnel input.

Coordinated Development: Coordinated development is a key criterion for evaluating high-quality development in the new era. It focuses on balanced development across different regions, sectors, and social groups. It not only involves coordination between urban and rural areas and among regions, but also includes balanced development between the economy, society, and the ecological environment. Therefore, adhering to coordinated development and integrating and balancing various interests are essential to achieving dynamic equilibrium between production and demand, improving overall economic efficiency, and enhancing developmental harmony. This study measures urban coordinated development through three indicators: regional coordination index, industrial coordination index, and financial coordination index.

Green Development: Green development is the defining characteristic of high-quality development. It emphasizes equal importance of economic growth and environmental protection, advocates low-carbon and eco-friendly development models, and pursues sustainable development. The concept of green development encourages the use of clean energy, promotes a circular economy, reduces pollution emissions, and protects the natural environment, aiming to achieve harmonious coexistence between humans and nature. This study evaluates urban green development based on three dimensions: waste emission index, pollution treatment index, and greening environment index.

Open Development: Openness is an essential path toward high-quality development. The concept of open development emphasizes combining "bringing in" with "going global", making full use of two markets (domestic and foreign) and two types of resources to promote a new pattern of all-round, multi-level, and wide-ranging openness. Upholding open development means deeply integrating into the global economy, actively adapting to the trend of economic globalization, and pursuing institutional, high-level opening-up, thereby promoting high-quality development through high-level openness. This study assesses urban open development through three indicators: foreign investment openness index, foreign trade openness index, and information openness index.

Shared Development: Shared development is the ultimate goal of high-quality development. As an essential requirement of socialism with Chinese characteristics, shared development focuses on addressing issues of social equity and justice. It emphasizes that the fruits of development should be shared by all people, ensuring that everyone benefits from economic progress, including improvements in people's livelihoods, promotion of fairness and justice, and narrowing income gaps. Therefore, fostering a sense of shared development and establishing more effective institutional arrangements are crucial to steadily advancing toward common prosperity. This study measures urban shared development through three indicators: completeness of educational facilities, completeness of cultural facilities, and completeness of medical facilities.

2.2 Construction of the High-Quality Economic Development Evaluation Model Based on Deep Learning

Model Applicability Evaluation. Deep learning is an artificial intelligence technology that uses learning sample data to extract feature representations of research objects. Its core mechanism involves artificial neural network models with multiple hidden layers [17,18], including Convolutional Neural Networks (CNNs), Recurrent Neural Networks (RNNs), Long Short-Term Memory networks (LSTMs), Generative Adversarial Networks (GANs), and Autoencoders.

An Autoencoder is a type of deep neural network capable of unsupervised learning, which can be regarded as a nonlinear and more robust alternative to Principal Component Analysis (PCA). An autoencoder consists of two symmetrical components: an encoder and a decoder, which reconstruct input data by ensuring consistency between input and output, while extracting latent features from hidden layers. The encoder first maps the data into higher-dimensional space to explore correlations among variables, then compresses them into lower-dimensional hidden features. The decoder then reconstructs the original data by decoding the learned hidden features in reverse [19]. Compared to PCA, which only handles linear relationships, autoencoders are capable of capturing both linear and nonlinear relationships in data.

The integration of deep learning and autoencoders not only enables the modeling of nonlinear relationships between data and evaluation outcomes but also significantly improves the accuracy of recognition algorithms, thereby optimizing traditional linear evaluation methods. Therefore, this study constructs a deep neural network model based on autoencoders to evaluate the level of high-quality economic development, as illustrated in Fig. 1.

Model Construction and Algorithm Implementation. This study uses Python's PyTorch library to construct a deep neural network for evaluating the level of high-quality economic development based on autoencoders. Initially, the input layer is set to include 20 normalized feature dimensions from the high-quality economic development dataset. Specifically, the input layer consists of 20 normalized indicators of high-quality economic development, with fully connected layers in the encoder's hidden layers. The first to fourth layers sequentially have 128, 512, 128, and 32 nodes, respectively, using the LeakyReLU activation function. The final layer has 1 node with the Tanh activation function, which outputs the data representation in the latent space. The decoder mirrors the structure of the encoder symmetrically, with hidden layers also being fully connected. It takes 1 node as the decoder input and has hidden layers with sequentially 32, 128, 512, and 128 nodes. The output layer of the decoder uses the

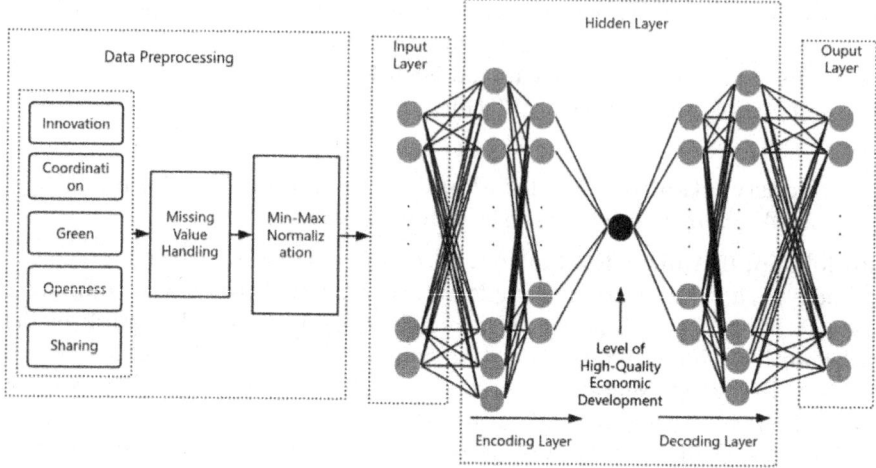

Fig. 1. Structure Diagram of a Deep Neural Network Model Based on Autoencoders.

same number of nodes as the input layer, reconstructing the high-quality economic development dataset after learning by the neural network. The maximum number of iterations is set to 200. Training will stop early if the validation loss does not improve over 30 consecutive epochs, thereby optimizing the training process for this deep neural network.

Activation Function

First, LeakyReLU (Leaky Rectified Linear Unit) is a simple and effective activation function that addresses the "dying neuron" problem of the traditional ReLU by introducing a small slope α in the negative half-axis. It is widely used in deep learning models, especially in convolutional neural networks (CNNs) and fully connected networks. Its mathematical formula is:

$$f(x) = \begin{cases} x & if \ x \geq 0 \\ \alpha x & if \ x < 0 \end{cases} \quad (1)$$

Here, α denotes a positive constant smaller than 1 (commonly set to 0.01 or another small value), referred to as the "leakage coefficient".

Second, Tanh Activation Function. Tanh (Hyperbolic Tangent) is a classical nonlinear activation function widely used in neural networks. It is a variant of the Sigmoid function and possesses symmetry as well as stronger nonlinearity. Its mathematical formula is:

$$f(x) = \frac{e^x - e^{-x}}{e^x + e^{-x}} \quad (2)$$

Here, e is the base of the natural logarithm.

Optimizer and Loss Function

The optimizer is a key component in the training process of deep neural networks, with the primary goal of effectively updating the model's weights to minimize the loss function. Among various optimization algorithms, Stochastic Gradient

Descent (SGD) is widely adopted in deep learning due to its simplicity and computational efficiency. Therefore, this study employs the SGD algorithm as the optimizer for the deep neural network. The SGD algorithm can be briefly expressed by the following formula:

$$w_0 \leftarrow w_0 + \alpha\left(y - h_w(x)\right) \tag{3}$$

$$w_1 \leftarrow w_1 + \alpha\left(y - h_w(x)\right) \tag{4}$$

In this context, w represents the weights between neurons in the neural network, α is the learning rate of the optimizer algorithm, which is set to 0.2 in this study, y denotes the expected output value of the neuron, and $hw(x)$ represents the actual output value of the neuron. In this study, the weights are calculated and updated sequentially for each data sample in the neural network training set until the loss function converges to the desired value.

The loss function in a deep neural network is used to quantify the difference between the model's predicted values and the actual (true) values. By minimizing the loss function, the deep neural network can learn the optimal mapping relationship from inputs to outputs. The Mean Squared Error (MSE) is a commonly used loss function for measuring the discrepancy between predicted and actual values in deep neural networks. Therefore, this study adopts MSE as the loss function for training. Its formula is briefly expressed as:

$$J_y(a) = \|Ya - b\|^2 \tag{5}$$

In this context, Y represents the matrix composed of all neurons in the input layer of the neural network, a denotes the weight matrices for each layer of the neural network, and b represents the expected output matrix.

Combining all the steps mentioned above, the parameters of the deep neural network model for evaluating high-quality economic development constructed in this study are summarized in Table 2.

Data Sources. This paper uses 3,124 data points from 284 prefecture-level cities across 30 provinces (autonomous regions and municipalities) during the period of 2011-2021 as research samples to scientifically measure the level of China's high-quality economic development. The specific sources and processing methods of the data are as follows: (1) Due to partial data missing in Tibet, the Hong Kong, Macao and Taiwan regions, these areas are excluded from the study; (2) The original data on imports and exports and foreign direct investment are denominated in US dollars, so they are converted into RMB based on the annual average exchange rate between the RMB and the US dollar for each year; (3) All research data are sourced from databases including Guoyan Network (DRCNet), China City Statistical Yearbook, CEINET Database, and CSMAR Database; (4) Missing data are supplemented through official sources such as the National Bureau of Statistics website, statistical yearbooks of various provinces and cities, and statistical bulletins on national economic and social development

Table 2. Neural Network Model Parameters for High-Quality Economic Development.

Parameter	Parameter Value
Number of Input Layer Nodes	20
Number of Hidden Layers	4
Number of Nodes per Hidden Layer	128, 512, 128, 32
Number of Output Layer Nodes	20
Hidden Layer Method	Fully Connected Layer
Activation Function	LeakyReLU, Tanh
Optimizer Selected	SGD
Optimizer Learning Rate	0.2
Optimizer Momentum Coefficient	0.9
Loss Function	MSE
Batch Size	17
Number of Iterations	Up to 200 (May stop early due to early stopping mechanism)

issued by individual cities. For a few indicators with missing values, interpolation or analogy-based estimation methods are applied.

Normalizing the original data not only provides a basis for subsequent data preprocessing and feature selection, but also improves the training speed and performance of the model when using gradient descent optimization methods. Therefore, before applying deep learning methods to measure the level of high-quality economic development, the original data are normalized, as shown in Fig. 2. According to Fig. 2, some features exhibit a very concentrated distribution, such as C1, C2, C3, and C15, indicating that they contain relatively less information and may contribute little to the model's discrimination ability. In contrast, feature C18 shows a more uniform distribution with a clear peak, indicating better discrimination capability and higher informational value.

3 Experimental Results and Analysis

3.1 Basic Analysis

Given that this study uses a sample of 284 cities from 2011 to 2021, the dataset is relatively large. To better illustrate the changes in high-quality economic development across different cities over time, a heatmap was employed to visually represent the disparities among cities and the trends within each city over time. As shown in Fig. 3, the horizontal axis represents the years, the vertical axis lists the various cities, and the color intensity reflects the level of high-quality economic development in each city.

Upon observing the color variations in the heatmap, several patterns can be identified: First, as time progresses, an increasing number of cities exhibit darker colors, indicating that the majority of cities experienced an improvement in their high-quality economic development index during this period. This trend

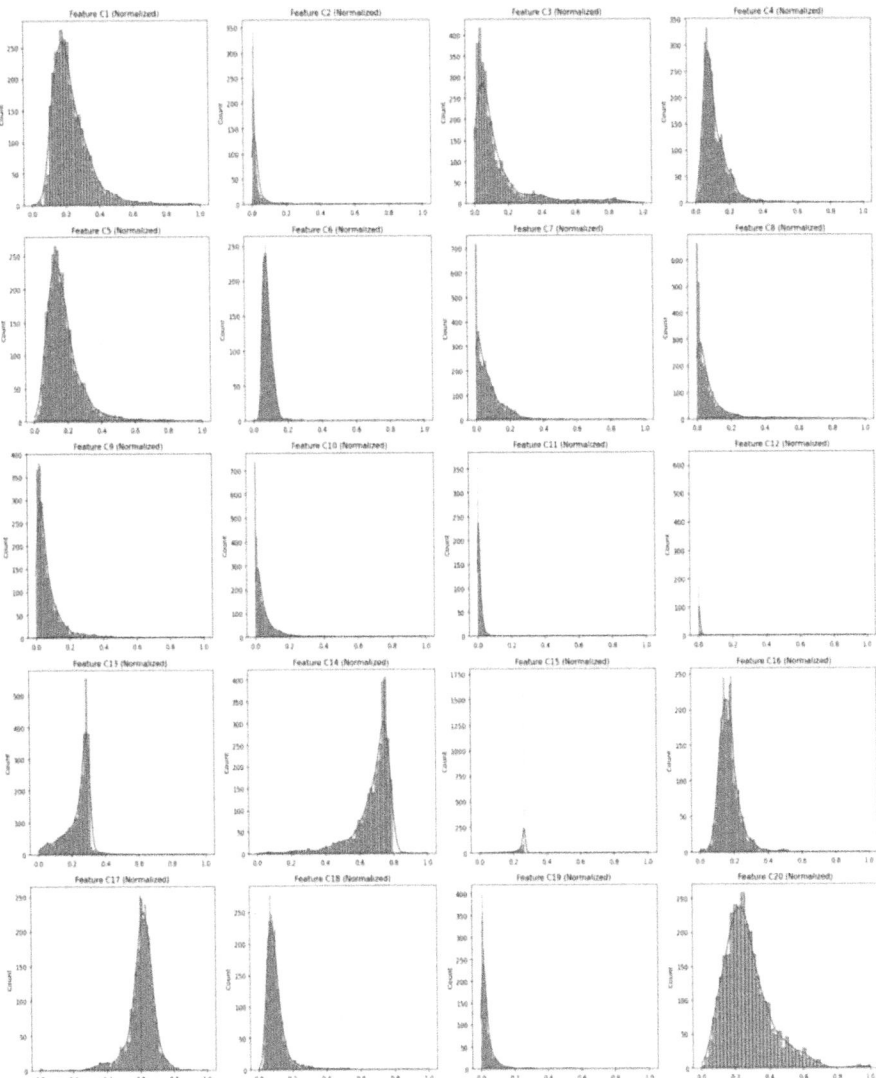

Fig. 2. Feature Visualization Map of the High-Quality Economic Development Dataset.

may be attributed to national policy support, adjustments in economic development strategies, and local reform measures implemented across regions. Second, in the earlier years, the color distribution was highly uneven, while in later years, the colors gradually became more uniform. The reduced variation in color intensity suggests that regional imbalances in economic development have somewhat improved, and the developmental gaps between regions have narrowed. Third, certain cities maintained consistently dark colors throughout the entire period, indicating sustained high levels of high-quality economic development.

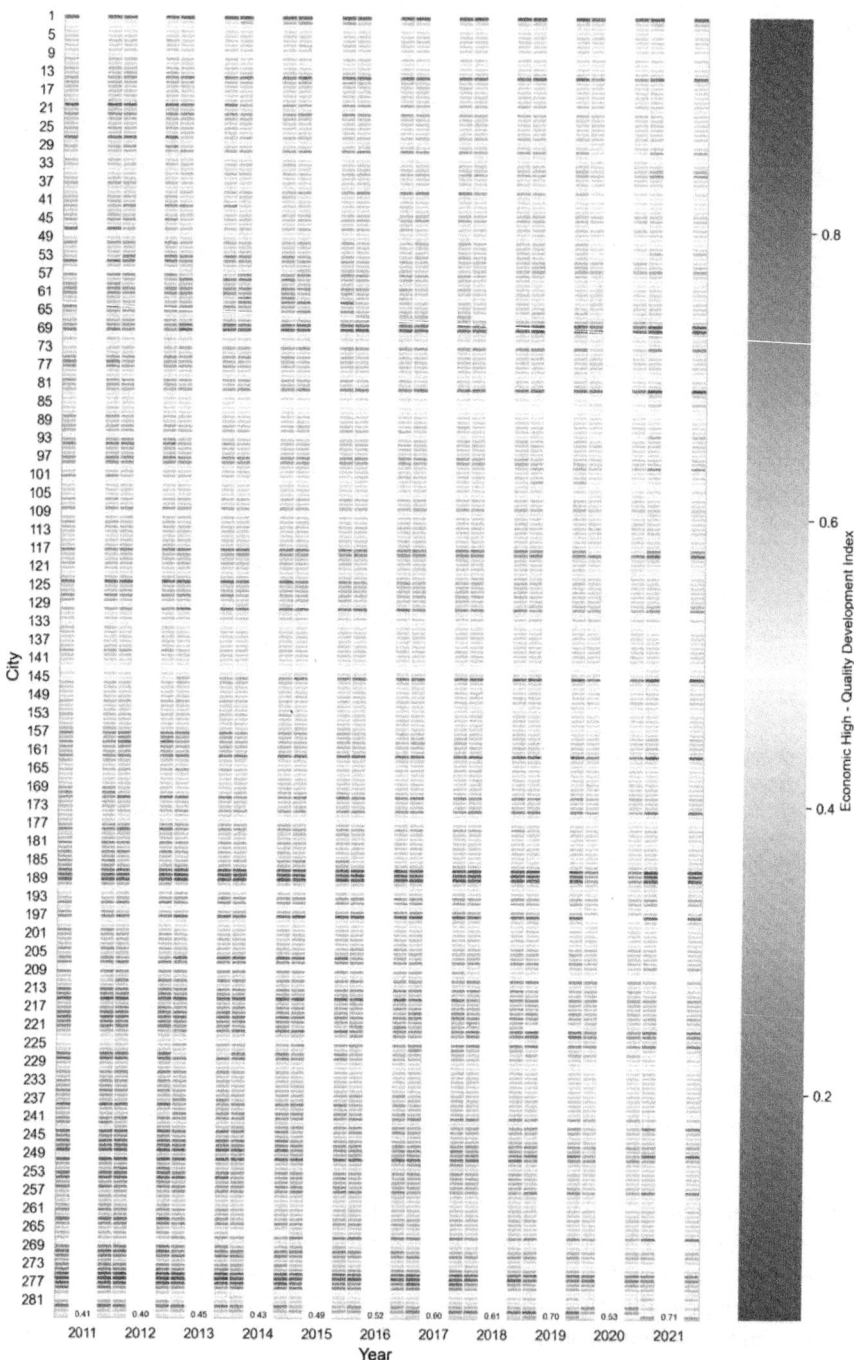

Fig. 3. Feature Visualization Map of the High-Quality Economic Development Dataset.

In contrast, some cities remained light-colored, reflecting ongoing development challenges they face.

In conclusion, although the high-quality economic development index of most cities shows an upward trend over time, there are still varying growth rates and fluctuations among different cities. This indicates the diversity and complexity inherent in China's urban economic development.

3.2 Comparative Analysis: Modern vs Traditional Methods

This Fig. 4 illustrates the mean values of high-quality economic development indices measured by three different methods from 2011 to 2021. It is evident that the deep learning approach (represented by a solid red line) performs best in assessing the high-quality economic development index, with its mean value significantly surpassing those calculated using the entropy weight method (orange dashed line) and the equal weighting assignment method (blue dotted line). Moreover, the deep learning method exhibits a consistently rising trend.

These findings suggest that compared to traditional methods (the entropy weight method and the equal weighting assignment method), the deep learning approach is more capable of capturing and reflecting the dynamic changes in high-quality economic development. This is particularly true when dealing with large and complex datasets, where deep learning demonstrates clear advantages. In contrast, traditional methods show limitations in capturing complex economic phenomena.

3.3 Kernel Density Estimation

Based on the above analysis, it is evident that the level of high-quality economic development in China exhibits significant spatiotemporal differentiation. To further investigate the dynamic evolution of disparities in urban high-quality economic development, this study employs Stata 16.0 to generate kernel density estimation maps of urban high-quality economic development levels. By analyzing the distribution patterns of the density curves, we can gain deeper insights into the evolutionary paths of high-quality economic development across cities. Figure 5 - 8 sequentially illustrate the dynamic evolutionary trends of high-quality economic development in the eastern, central, western, and northeastern regions of China, respectively.

This Fig. 5 illustrates the trend of high-quality economic development in the eastern region from 2011 to 2021. Observing the changes in peak values, it can be seen that in 2011, the curve's peak was relatively low and positioned towards the left. By 2014, the peak reached its highest point and remained relatively leftmost among the years observed. In 2018, the peak decreased and shifted rightward; by 2021, the peak further declined with a continued rightward shift. This indicates that over time, the concentrated position of the high-quality economic development index is gradually moving rightward, suggesting an overall increasing trend in the index values. Looking at the changes in the shape of the curves, from 2011 to 2014, the curve became taller and narrower, indicating that data points were

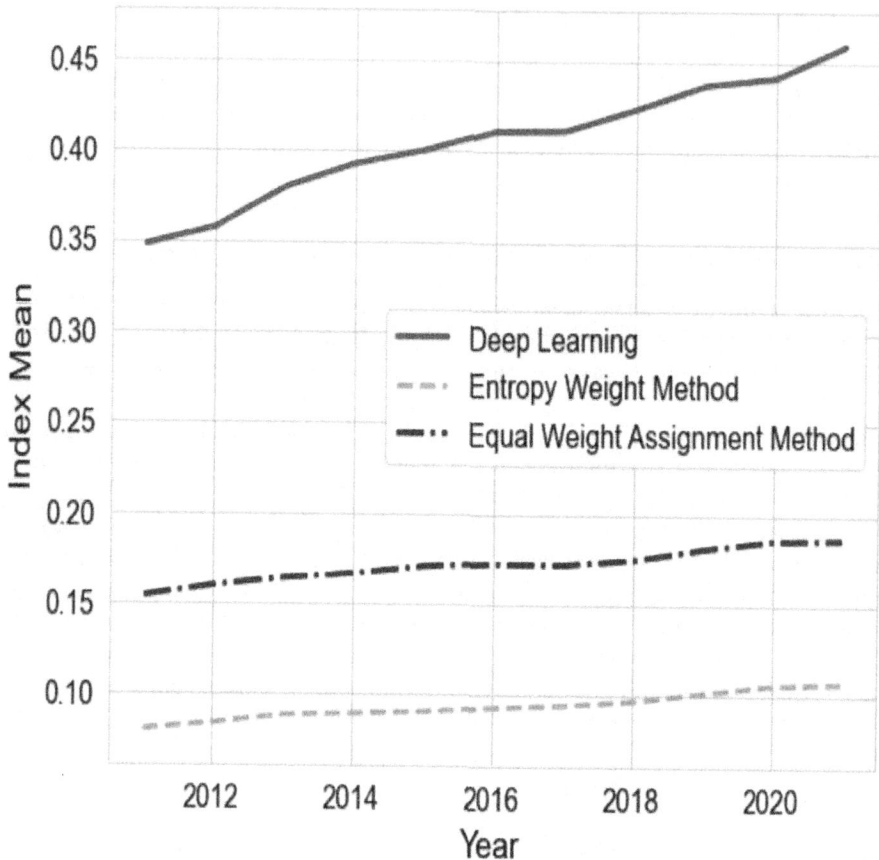

Fig. 4. Comparative Analysis.(Color figure online)

becoming more concentrated around a specific index value. From 2014 to 2021, the curve gradually became shorter and wider, reflecting an increasing dispersion of data points and growing differences in various index values. This suggests that while some areas may have experienced rapid development, others are lagging behind, highlighting disparities within the region. These findings indicate that over the past decade, the eastern region has made significant progress in terms of high-quality economic development. However, attention should still be paid to balanced development within the region to address these disparities.

Figure 6 illustrates the trend of high-quality economic development in the central region from 2011 to 2021. Observing the changes in peak values, it can be seen that in 2011, the curve's peak was relatively low and positioned towards the left. By 2014, the peak had risen and shifted rightward; in 2018, the peak further increased and continued to move rightward. In 2021, the peak reached its highest point and was located at the far right. This indicates that over time, the concentrated position of the high-quality economic development index for the

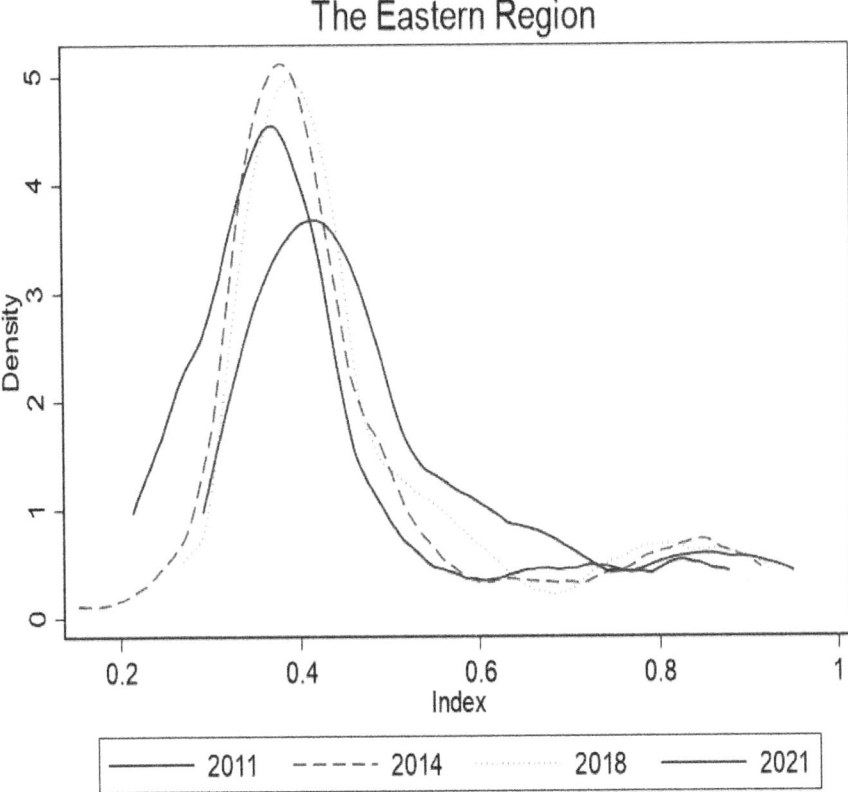

Fig. 5. The Evolutionary Trend of the High-Quality Economic Development Index in the Eastern Region.

central region is continuously moving rightward, suggesting a continuous increase in overall index values. Examining the changes in the shape of the curves, from 2011 to 2021, the curve gradually became taller and narrower, indicating that data points were becoming increasingly concentrated around specific index values. This suggests that the dispersion of the high-quality economic development index distribution in the central region has been gradually decreasing, meaning that the differences in high-quality economic development levels within the region have been narrowing. These findings indicate that from 2011 to 2021, the central region experienced a significant overall improvement in high-quality economic development, with an increasing balance in development across the region.

Figure 7 illustrates the trend of high-quality economic development in the western region from 2011 to 2021. Observing the changes in peak values, it can be seen that in 2011, the curve's peak was relatively positioned towards the left. From 2014 to 2018, the peak gradually increased and continued to shift rightward. This indicates that over time, the concentrated position of the high-quality

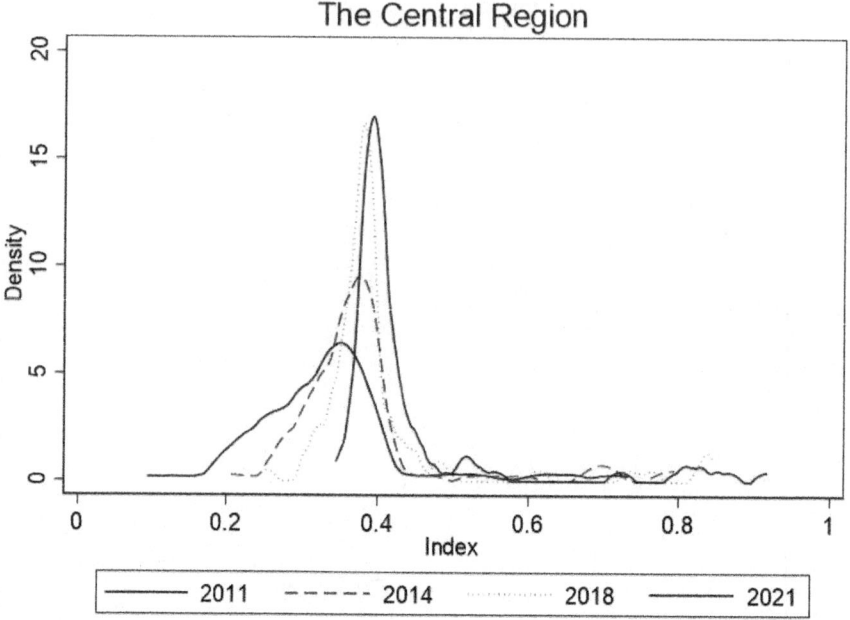

Fig. 6. The Evolutionary Trend of the High-Quality Economic Development Index in the Central Region.

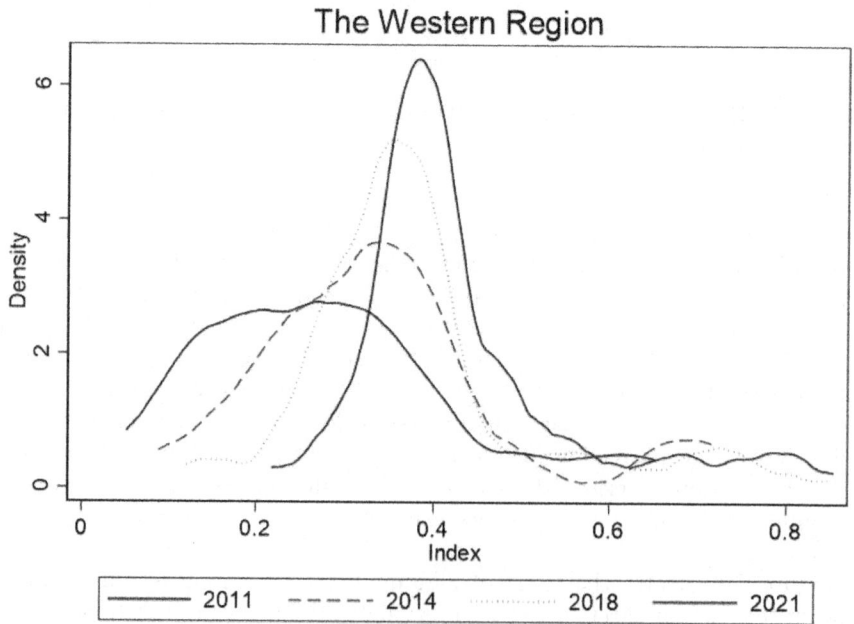

Fig. 7. The Evolutionary Trend of the High-Quality Economic Development Index in the Western Region.

economic development index for the western region is continuously moving rightward, suggesting a continuous increase in overall index values. Examining the changes in the shape of the curves, from 2011 to 2021, the curve initially became taller and narrower, indicating that data points were becoming increasingly concentrated around specific index values. This suggests that the differences in high-quality economic development levels within the region have been gradually decreasing, implying an enhancement in the balance of development across the region. However, in 2021 compared to 2018, there is a slight widening trend in the curve, indicating a slight increase in the dispersion of data points. This may suggest that some variations in development levels are beginning to emerge within the region. These findings indicate that from 2011 to 2021, the western region experienced a significant overall improvement in high-quality economic development. Nevertheless, attention should be paid to coordinated development within the region.

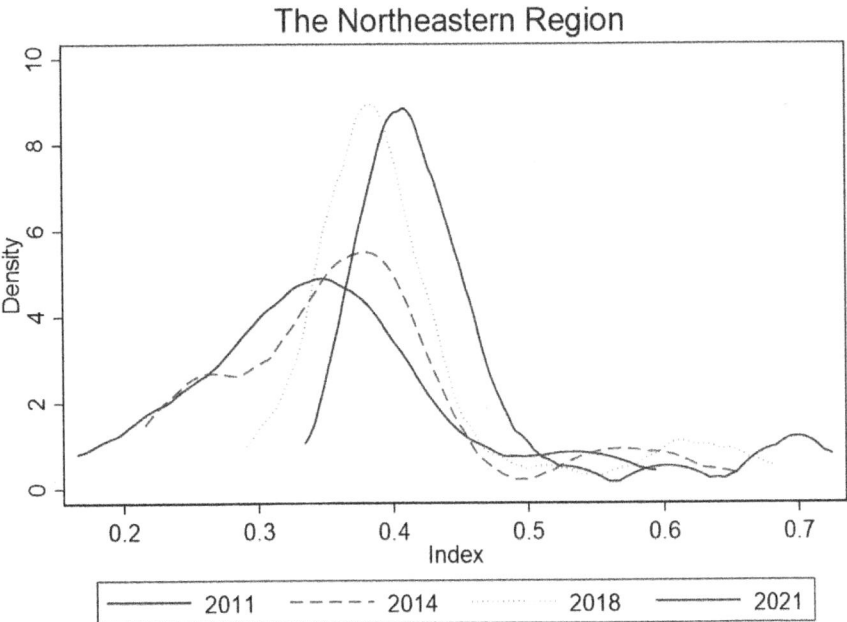

Fig. 8. The Evolutionary Trend of the High-Quality Economic Development Index in the Northeastern Region Region.

Figure 8 illustrates the trend of high-quality economic development in the northeastern region from 2011 to 2021. Observing the changes in peak values, it can be seen that from 2011 to 2018, the curve's peak increased and shifted rightward; however, in 2021, the peak decreased compared to 2018 but remained on the right side. This indicates that although the average level of high-quality economic development continued to rise over time, the concentration at the highest peak decreased, showing that different areas within the region began to develop at varying speeds. Examining the changes in the shape of the curves, from 2011 to 2018, the curve gradually became taller and narrower, indicating that data points were becoming increasingly concentrated around specific index values, suggesting an enhancement in the balance of development within the region. From 2018 to 2021, the curve widened and the peak lowered, indicating an increase in the dispersion of data points. This suggests that some areas experienced faster development while others progressed more slowly. These findings indicate that from 2011 to 2021, the level of high-quality economic development in the northeastern region showed significant improvement, but attention should be paid to the coordination of development within the region.

In summary, from 2011 to 2021, the high-quality economic development indices of the four major regions (eastern, central, western, and northeastern) all showed varying degrees of improvement, indicating that the overall levels of high-quality economic development across these regions have advanced. However, attention should still be paid to issues related to the balance and coordination of development within each region.

3.4 Markov Chain

To further analyze the spatial evolution trends of high-quality economic development in different regions, this study employed both traditional Markov chains and spatial Markov chains to predict the probabilities of regions transitioning between different levels of high-quality economic development. The quartile method was used to categorize the 284 cities'high-quality economic development indices into four grades: low, lower-middle, upper-middle, and high, based on their values in ascending order. The results are shown in Table 3.

Table 3. Traditional Markov Transition Probability Matrix of China's High-Quality Economic Development from 2011 to 2021.

Category	Low	Lower-Middle	Upper-Middle	High
Low	0.6583	0.3417	0.0000	0.0000
Lower-Middle	0.0129	0.9669	0.0198	0.0005
Lower-Middle	0.0000	0.0438	0.8418	0.1145
Lower-Middle	0.0000	0.0078	0.1484	0.8438

According to Table 3, the following conclusions can be drawn: First, the maximum probability on the diagonal is 0.9669, while the minimum is 0.6583, both of which are greater than the off-diagonal probabilities. This indicates that China's high-quality economic development exhibits significant stability in maintaining its current state across different levels. Second, when economic development is at either low or high levels, the probability of remaining in the same state in the next period is extremely high. This suggests the existence of a "club phenomenon" in economic development low-level regions tend to cluster and persist at low levels, while high-level regions similarly maintain their status, making it difficult for regions to transition between different "clubs". Third, the transition probabilities from "low" to "medium-high" or "high", from "medium-low" to "high", and from "medium-high" to "low" are all zero. This implies that, between consecutive years, cities'high-quality economic development cannot achieve "leapfrog" progress along these cross-level transition paths. Although some adjacent-level transitions may occur, overall, upward mobility across tiers remains highly challenging.

Table 4. Traditional Markov Transition Probability Matrix of China's High-Quality Economic Development from 2011 to 2021.

Spatial Lag	Type	Low	Medium-Low	Medium-High	High
Low	Low	0.7308	0.2692	0	0
	Medium-Low	0.0342	0.9410	0.0248	0
	Medium-High	0	0.0247	0.8148	0.1605
	High	0	0	0.2917	0.7083
Medium-Low	Low	0.5238	0.4762	0	0
	Medium-Low	0.0093	0.9728	0.0180	0
	Medium-High	0	0.0488	0.8488	0.1024
	High	0	0	0.1279	0.8721
Medium-High	Low	0	0	0	0
	Medium-Low	0	0.8235	0.1176	0.0588
	Medium-High	0	0.0909	0.9091	0
	High	0	0.0556	0.0556	0.8889
High	Low	0	0	0	0
	Medium-Low	0	0	0	0
	Medium-High	0	0	0	0
	High	0	0	0	0

By further incorporating the neighborhood environment of cities (where adjacent cities are coded as 1 and non-adjacent as 0), we obtain the spatial Markov transition probability matrix presented in Table 4.

First, China's high-quality economic development demonstrates significant spatial spillover effects. When neighboring cities advance from low to medium-high development levels, upward transition probabilities increase for low-, medium-low-, and medium-high-level cities. This suggests mutual reinforcement effects among regions with relatively small development gaps, where higher-level neighbors likely facilitate peripheral development through technology diffusion, industrial transfer, and infrastructure connectivity - indicative of "trickle-down effects" or "spread effects".

Second, however, when neighbors reach high development levels, compared to medium-low-level neighbors, transition probability from low to medium-low level cities decreases, and medium-high-level cities show reduced upward mobility to high level. This implies that excessive development gaps may nullify positive spillovers, as high-level regions potentially absorb quality resources through "polarization effects", creating development bottlenecks that prevent adjacent medium-high-level cities from achieving further advancement.

4 Conclusions and Future Works

4.1 Conclusions

This study constructs an evaluation model of high-quality economic development in China based on deep learning methods, and employs kernel density estimation and Markov chains to explore the spatiotemporal evolution and spatial transition patterns of high-quality economic development. The main findings are as follows: First, based on the color variation in the heatmap, it is observed that over time, the high-quality economic development index has generally shown an upward trend across most cities. However, there still exist varying growth rates and fluctuations among different cities. Second, by comparing the results measured using traditional methods-the entropy weight method (EWM) and the equal weighting method (EWM)-it is found that the deep learning (DL)-based measurement of the high-quality economic development index outperforms conventional linear evaluation approaches in terms of accuracy and performance. Third, through kernel density estimation mapping, this study reveals the dynamic evolutionary trends of regional high-quality economic development. Over the past decade, all four major regions-eastern, central, western, and northeastern China-have experienced varying degrees of improvement in their high-quality economic development indices. However, attention should still be paid to issues of internal regional coordination and balanced development. Fourth, the analysis based on the Markov chain reveals the spatial evolution trends of high-quality economic development in China. It shows that China's high-quality economic development is significantly influenced by spatial spillover effects from neighboring regions. However, the direction and intensity of these effects depend on the economic development gaps between regions. When the development gap is small, there is a clear mutual promotion effect. In contrast, when the gap is large, the driving effect of high-level regions on low-level regions becomes limited.

4.2 Policy Recommendations

This paper constructs an evaluation model of China's high-quality economic development based on deep learning method, and combines kernel density estimation and Markov chain method to reveal its spatial-temporal dynamic evolution characteristics, which provides the following implications for policy making:

First, the results show that the level of high-quality economic development in most cities shows an upward trend, but there are significant differences between regions. This result indicates that differentiated policy design should be paid attention to in the process of promoting the overall high-quality development of the country, and "one size fits all" should be avoided. According to the development stage, industrial structure and innovation capacity of each region, the central and local governments can formulate hierarchical and classified guiding policies to strengthen the coordinated development mechanism among regions.

Second, the high-quality development index measured by the deep learning model is better than the traditional linear method, indicating that the introduction of nonlinear modeling tools can more accurately identify the key drivers in economic development. Therefore, it is suggested that relevant departments introduce artificial intelligence and big data analysis technology into the future performance evaluation system, improve the scientific and forward-looking policy evaluation, and realize the transformation from experience-driven to data-driven governance mode.

Third, the results of kernel density estimation show that the eastern, central, western and northeastern regions all show different degrees of development and improvement trends, but the internal coordination of the regions still needs to be strengthened. Therefore, in the future, we should further optimize the strategic layout of regional development, promote infrastructure connectivity and factor market integration, narrow the gap within regions, and enhance the endogenous growth capacity of low-level regions.

Fourthly, Markov chain analysis reveals the existence of spatial spillover effect, and its direction is affected by the economic development gap between regions. Therefore, policy makers should pay attention to the construction of regional collaborative innovation networks and encourage high-level regions to drive the development of surrounding areas through technology diffusion and industrial transfer. At the same time, the targeted support mechanism for underdeveloped regions should be improved to prevent the Matthew effect of "the strong always getting stronger and the weak getting weaker" from aggravating the imbalance of regional development.

4.3 Future Works

First, in the future research work, deep learning methods and traditional linear methods will be combined to further measure the status of high-quality economic development. The task of deep learning is to extract key features from complex economic data, while traditional linear methods build intuitive and easily interpretable models based on these features. This synergy makes the measurement of high-quality economic development more comprehensive and accurate.

Second, in future research, we will use SHAP and LIME methods to explain the decision-making process of the model, and introduce a causal reasoning framework to evaluate the actual impact of different policies on urban economic development, aiming to develop a model system with both high prediction accuracy and good interpretability, so as to better support urban planning and policy making.

Acknowledgment. This work was supported by the National Nature Science Foundation of China (No. 62462054), the Jiangxi Provincial Natural Science Foundation Project of China (No. 20232BAB202007), and the Jiangxi Province Higher Educational Reform Research Project of China (No. JXJG-23-17-25).

Disclosure of Interests. The authors declare no conflicts of interest.

References

1. Barahona, R., Maria, L.: Deep learning for sentiment analysis. Language & Linguistics Compass **10**(12), 205–212 (2016)
2. Wang, H., et al.: Research on urban base land price evaluation model using deep learning. China Land Sci. **32**(9), 59–65 (2018)
3. Bei, J.: Research on high-quality development from an economic perspective. China Industr. Econo. **4**, 5–18 (2018)
4. Pu, X., Jarko, F.: Research on the optimization mechanism of the driving structure for china's high-quality economic development. J. Northwest Univ. (Philosophy Soc. Sci. Edn.) **1**, 113–118 (2018)
5. Ren, B.: Judgment criteria, determinants, and implementation approaches for China s high-quality development in the new era. Reform **4**, 5–16 (2018)
6. Wang, X., Xu, X.: Spatiotemporal evolution and regional disparities of high-quality economic development in the yangtze river economic belt. Economic Geography **40**(03), 5–15 (2020)
7. Wang, Z., Yang, L., Yin, J., et al.: Assessment and prediction of environmental sustainability in China based on a modified ecological footprint model. Res. Conserv. Recycling, 301–313 (2017)
8. Fang, S., Xiao, Q.: Study on regional ecological welfare performance levels and their spatial effects in China. China Population. Res. Environ. **29**(03), 1–10 (2019)
9. Wu, K., Shi, J., Yang, T.: Has energy efficiency performance improved in china? non-energy sectors evidence from sequenced hybrid energy use tables. Energy Econo. **67**(09), 169–181 (2017)
10. Badinger, H.: Output volatility and economic growth. Econ. Lett. **106**(01), 15–18 (2010)
11. Tian, X., Zhang, X., Zhou, Y., Yu, X.: Regional income inequality in china revisited: a perspective from club convergence. Econ. Model. **56**, 50–58 (2016)
12. Ma, R., Hui, L., Hongwei, W., et al.: Research on the evaluation index system and measurement of china's regional high-quality economic development. China Soft Sci. **07**, 60–67 (2019)
13. Guo, Y., Fan, B., Long, J.: Research on actual measurement and spatiotemporal evolution characteristics of regional high-quality development in China. J. Quantitative Techn. Econo. **37**(10), 118–132 (2020)
14. Pan, J., Zheng, H.: Spatiotemporal evolution characteristics of regional economic high-quality development differences. Stat. Deci. **37**(24), 88–92 (2021)
15. Chenhui, D., Ze, T., Xiaoming, S., et al.: Research on regional economic high-quality development in china under the new development philosophy c level measurement, spatiotemporal differentiation and dynamic evolution. Techn. Manage. Res. **12**, 3–9 (2022)

16. Su, S., Chen, X.: The impact of the belt and road initiative on the high-quality economic development of Chinese node cities along the route an examination based on the mediating effect of improved infrastructure construction. J. Qinghai National. Univ. (Soc. Sci. Edn.) **51**(01), 125–136 (2025)
17. LeCun, Y., Bengio, Y., Hinton, G.: Deep learning. Nature **521**(7553), 436–444 (2015)
18. Liu, X., Wu, Z., Luo, R., et al.: Urban fringe identification based on multi-source data and deep learning. Geograph. Res. **39**(02), 243 C256 (2020)
19. P. Autoencoders, B.: unsupervised learning and deep architectures. In: JMLR: Workshop and Conference Proceedings, vol. 27, p. 37 C50 (2012)

TLAQ: Enhanced Big Healthcare Data Analytics via Lazy Aggregations

Alfredo Cuzzocrea[1,2(✉)], Islam Belmerabet[1], and Abderraouf Hafsaoui[1]

[1] iDEA Lab, University of Calabria, Rende, Italy
{alfredo.cuzzocrea,islam.belmerabet,
abderraouf.hafsaoui}@unical.it
[2] Department of Computer Science, University of Paris City, Paris, France

Abstract. This paper introduces *Tree-Like Analytical Queries* (TLAQ) model for supporting *enhanced big healthcare data analytics* via an innovative concept, the so-called *lazy aggregations*. Given a *hierarchical tree-like aggregate query*, which fully supports advanced big data analytics tools, according to the lazy aggregation paradigm, data ranges of two parent-child nodes *do not* satisfy the *containment relation*, thus opening the door to detailed implementations of target *medical investigation processes* (e.g., in the context of epidemiological research). The latter innovation turns to be extremely useful in modern big healthcare data analytics, as proofed in this paper. We finally provide a comprehensive case study about the potentialities of the TLAQ analytical model on top of a real-life case study deriving from a reference EU H2020 research project.

Keywords: Big Healthcare Data · Big Healthcare Data Analytics · Multidimensional Big Healthcare Data Analytics

1 Introduction

In this paper, we introduce the *Tree-Like Analytical Queries* (TLAQ) model as a novel approach to enhance *big healthcare data analytics* (e.g., [1, 2]), being an innovative approach proposed within the context of the *EU H2020 project QUALITOP* [3]. The TLAQ model leverages an innovative paradigm referred to as *"lazy aggregations"* to strictly follow *a typical medical investigation process* (e.g., [4]) and optimize the efficiency and *granularity* of data processing in the context of large-scale healthcare datasets. The model aims to address key challenges associated with processing and analyzing massive volumes of healthcare data by supporting advanced analytics tools through the application of *hierarchical, tree-like aggregate queries*, which enable more refined and *targeted* data analysis, being the latter particularly crucial in healthcare.

According to the lazy aggregation paradigm, the data ranges of parent-child nodes within the tree structure do not necessarily adhere to the traditional *containment relationship*, as is commonly seen in conventional *data aggregation techniques* (e.g., [5]).

This research has been made in the context of the Excellence Chair in Big Data Management and Analytics at University of Paris City, Paris, France

This unique characteristic allows for a greater level of flexibility and precision when handling complex datasets, providing the opportunity for more nuanced and detailed implementations of medical investigations, as said. Specifically, this feature enables the model to support more sophisticated *disease analysis*, where data distributions and relationships are often *non-linear and interdependent*. By enabling partial and incremental aggregation of data, the TLAQ model allows us to examine *finer-grained* trends and correlations that may have been obscured in more traditional data analysis methods.

The proposed TLAQ model is implemented within the core layer of the *QUALITOP Federated Big Data Analytics Learning System* (QFLS) [6, 7], a core Cloud-based framework of the QUALITOP big data ecosystem.

While the paper mostly focuses on the theoretical model of TLAQ, to further illustrate the practical potential of the surrounded analytical model, we present a comprehensive case study that utilizes a real-life healthcare dataset, specifically, the *IMMUCARE* dataset coming from the QUALITOP project. This dataset, which is *federated* across QFLS by largely marrying *data federation paradigms* (e.g., [8, 9]), serves as an ideal testbed for demonstrating the capabilities of our model. By providing a detailed running example of how the TLAQ model functions within the execution framework of QFLS, we showcase its effectiveness in real-world scenarios. The case study highlights how the model facilitates efficient data querying, aggregation, and analysis, even in the presence of *distributed* and *heterogenous big data sources* (e.g., [10–12]). It also underscores the practical applicability of the TLAQ model in supporting complex healthcare research, and it demonstrates how its implementation can enhance both the scope and depth of medical investigations conducted on large, *federated healthcare datasets* (e.g., [13]).

2 QFLS System: A Brief Overview

In this Section, we present a concise overview of the QFLS system, by detailing its structural components and core functionalities. Figure 1 depicts the architecture of QFLS in the context of the QUALITOP research project.

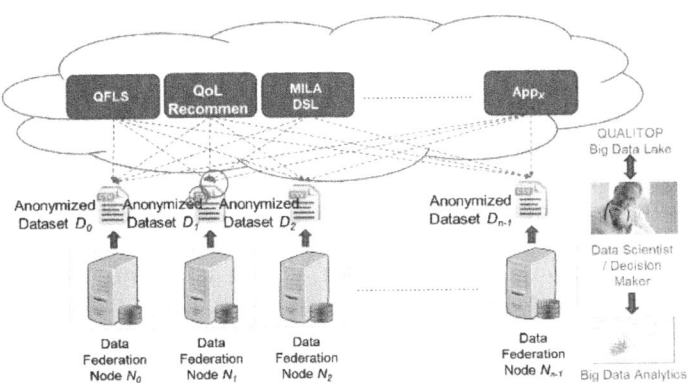

Fig. 1. Architecture of the QFLS System.

As shown in Fig. 1, QFLS relies on the *QUALITOP Big Data Lake* [14], which is overall composed by the *Data Federation* of the participant nodes (e.g., hospitals, medical centers, etc.) providing *anonymized datasets* plus several analytical applications, including: (*i*) *MILA DSL* [3], a component for supporting interoperability among different, heterogenous medical data sources; (*ii*) *QoL Recommender* [3], a component for deriving medical recommendations about quality-of-life based on *Machine Learning* algorithms (e.g., [15–17]). Similarly, the framework is open to the inclusion of other applications in the big data lake, according to well-understood *Cloud Computing* predicates (e.g., [18, 19]).

The QFLS system is a comprehensive big data analytics framework designed to support both exploratory and predictive data analysis, with privacy-preserving features. Its primary objective is to facilitate the extraction of actionable insights from large-scale, anonymized datasets within a *Federated Learning environment* (e.g., [20, 21]), on the basis of the TLAQ analytical model. QFLS provides an intuitive and flexible environment that empowers users to design, execute, and manage complex data analysis workflows efficiently. This way, the overall proposal is suitable for specialized contexts, such as the healthcare context, like in the QUALITOP research project.

The QFLS architecture consists of two primary modules: (*i*) the *Data Ingestion* component, and (*ii*) the *Data Analytics* component. While the Data Ingestion component is responsible for the acquisition and pre-processing of data, the primary focus of this study is the Data Analytics component, which retains a prominent interest in advanced big data analytics tools.

The Data Analytics component is designed to enable the analysis of anonymized datasets distributed across multiple federated nodes. Importantly, QFLS maintains strict data privacy protocols, ensuring that raw datasets are never directly shared between nodes. Instead, the system retrieves aggregate analytical outcomes from the respective federated nodes, thereby supporting large-scale data analytics in a *privacy-preserving manner* (e.g., [22]). This privacy-centric approach not only safeguards sensitive information (e.g., [23–25]), but also leverages partial analytical outputs from distributed datasets to enhance decision-making processes and guide evidence-based treatment recommendations.

In the following, we show an example TLAQ built on top of the main table *Tumour* of the real-life dataset *Simulacrum* [26], whose relational schema is shown in Fig. 2. *Simulacrum* realizes a synthetic database that stores data about patients with cancer, particularly it stores data about cancer patients' *medical histories, clinical trials, therapies*, and various other aspects of *cancer care*. Although being synthetic, *Simulacrum* maintains most of the properties of the original data with a high degree of accuracy. *Simulacrum* is a collection of linked data tables, organized in two main groups: (*i*) Cancer Registration tables (SIM_AV); (*ii*) Systemic Anti-Cancer Therapy tables (SIM_SACT).

Figure 3 shows the example TLAQ $TLAQ_{Simulacrum}$ on top of the table *Tumour* (see Fig. 2). As shown in Fig. 3, at root node N_0, $TLAQ_{Simulacrum}$ queries all the *Male* patients, and retrieves the COUNT aggregation over them. Then, at the first level, at node $N_{1.1}$, $TLAQ_{Simulacrum}$ queries, within the precedent root node dataset, male patients of age *over 40 years old*. At the second level, at node $N_{2.1}$, $TLAQ_{Simulacrum}$ queries, within the precedent $N_{1.1}$ dataset, male patients of age over 40 years old and having *ICD10 site*

Fig. 2. The Dataset *Simulacrum*.

code equal to C440. The rest of node queries of the target TLAQ $TLAQ_{Simulacrum}$ follow a similar (TLAQ model) paradigm.

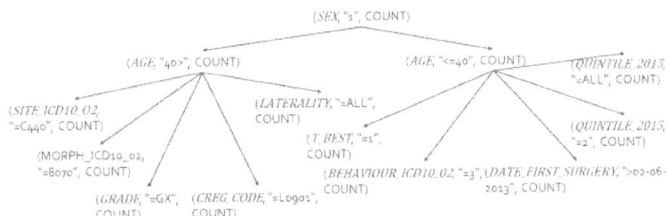

Fig. 3. Sample TLAQ $TLAQ_{Simulacrum}$.

By executing $TLAQ_{Simulacrum}$, we retrieve a tree-like data structure such that every node links the corresponding dataset analytics. In turns, on top of these latter datasets, it is possible to generate various kinds of *visual big data analytics structures* for deep meaning and knowledge insights. Looking at the specific case, $TLAQ_{Simulacrum}$ provides insights into *age-related trends* in cancer diagnosis, treatment, and outcomes, highlighting differences in *cancer prevalence*, *surgery timing*, and *severity across age groups*. Additionally, it can reveal disparities in cancer care based on *socio-economic or geographical factors*, thus helping into refining targeted treatment strategies.

To optimize the analytical process, QFLS employs *data flattening techniques* to condense multidimensional, multi-level hierarchical data, originally structured within TLAQs, into *multidimensional dimensional summaries*. Figure 4 shows, as an instance, the case of a two-dimensional summary, for the sake of simplicity. This transformation simplifies complex datasets, enabling more efficient *cross-query* comparisons and aggregate analysis over summary data. The flattened data structure supports comprehensive insights across diverse analytical dimensions, promoting nuanced decision-making and detailed cross-sectional evaluations.

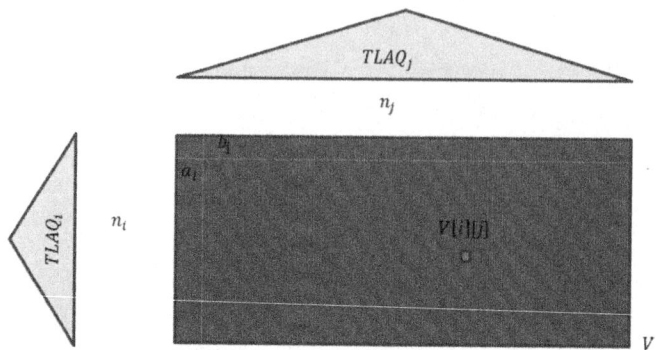

Fig. 4. Two-Dimensional Summary (TLAQ Model).

Consider the two-dimensional case depicted in Fig. 4. Here, from the leaf levels of the target TLAQs $TLAQ_i$ and $TLAQ_j$, the proposed methodology introduces an *approximate computation* of the summary V, via computing every data cell in V, denoted by $V[i][j]$, as follows:

$$V[i][j] \approx \left\lfloor \frac{a_i}{\sum_{i=0}^{|n_i|-1} a_i} \times \frac{b_j}{\sum_{j=0}^{|n_j|-1} b_j} \times \frac{\sum_{i=0}^{|n_i|-1} a_i + \sum_{j=0}^{|n_j|-1} b_j}{n_i + n_j} \right\rfloor \quad (1)$$

wherein: (*i*) n_i is the number of leaf-level nodes of $TLAQ_i$; (*ii*) n_j is the number of leaf-level nodes of $TLAQ_j$; (*iii*) a_i is the value of the leaf-level node of $TLAQ_i$ at position i; (*iv*) b_j is the value of the leaf-level node of $TLAQ_j$ at position j.

As in Eq. (1), the meaning of the approximation formula introduced by the TLAQ model consists in a *proportional estimation* of the contribution due to each data cell $V[i][j]$ with respect to the overall contributions of the *"frontiers"* given by the leaf levels of the target TLAQs. Overall, this methodology aggregates multidimensional data by: (*i*) combining proportions from the leaf levels of the target TLAQs; (*ii*) balancing their contributions based on the width of the leaf level of the target TLAQs; (*iii*) normalizing and averaging data as to create a concise summary in a multidimensional fashion.

The *Analytics and Predictive Analytics Environment* (APAE) module represents the culmination of the QFLS analytical workflow. It features an integrated dashboard designed to support data-driven decision-making through advanced predictive models. The dashboard aggregates a suite of analytical metrics, including: (*i*) *Aggregates*; (*ii*) *Averages*; (*iii*) *Standard Deviations*; (*iv*) *2D Clustering*; (*v*) *Icicles*; (*vi*) *Confusion Matrix*; (*vii*) *Pearson Correlation* [27]; (*viii*) *Spearman Correlation* [28]; (*ix*) *Outlier Detection*.

These metrics derive from the analytical outputs generated by TLAQ executions, facilitating the development of predictive treatment recommendations. Analytical outcomes are exportable in PDF format, enhancing their utility for reporting and strategic decision-making. Figure 5 depicts plots of several analytical metrics resulting from the QFLS TLAQs cross-analysis.

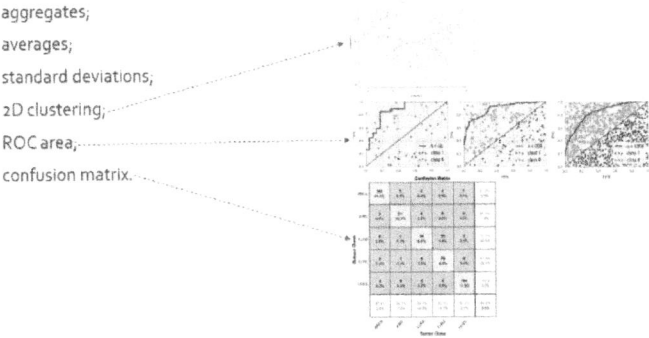

Fig. 5. Outcome Result Examples of the QFLS Analytical Metrics.

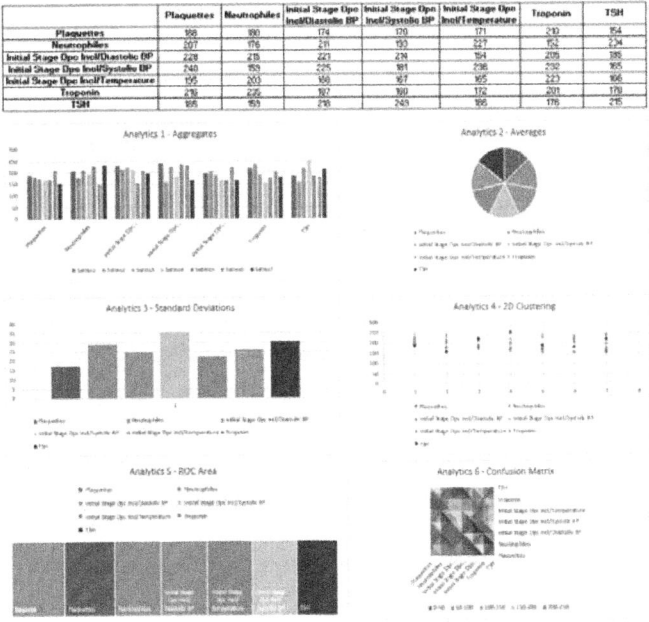

Fig. 6. Analytical Dashboard of the QFLS TLAQ Model Metrics.

Such an approach allows for the effective aggregation and comparison of data across multiple analytical queries. Flattening reduces the complexity of multidimensional information, facilitating a streamlined and unified analysis. It thus enables us to derive comprehensive insights from diverse data sources, supporting more nuanced decision-making and in-depth cross-sectional analysis.

In Fig. 6, we show a dashboard representing the analytical result metrics obtained based on the performed predictive models of QFLS TLAQ model.

The QFLS represents an innovative framework for conducting sophisticated analytics on extensive anonymized datasets. By leveraging a federated architecture and advanced big data processing methodologies, the system facilitates robust data exploration, predictive analytics, and evidence-based treatment recommendations. The integration of specialized modules, including the *Data Federation Discoverer*, *Dataset Anonymization Analysis*, and *Analytics Environment*, ensures that data analysts can securely and effectively manage, query, and analyze heterogeneous datasets. This architecture not only enhances *data-driven decision-making* within healthcare but also extends its applicability to a broad range of domains requiring secure, large-scale data analytics.

3 The Tree-Like Analytical Query Model

In this Section, we provide a formal description of the so-called *Tree-Like Analytical Query* model.

Tree-Like Analytical Query model is a *hierarchical data analysis model* designed to facilitate *enhanced big data analytics* over large-scale datasets, particularly in domains such as *healthcare*. TLAQ organizes analytical queries into a *structured tree-based format*, which allows for progressive and controlled data exploration. Moreover, the TLAQ model can be formally defined as follows:

Given a dataset D, a TLAQ model defined over D is an *n-ary tree* T, defined as follows:

$$T : \langle N, D, A, AP, \langle Cond \rangle \rangle \tag{2}$$

where N is the name of the *federated node*; D is the given dataset in N; A is the *selected attribute* in D; AP represents a *user-defined aggregation predicate* on A; $Cond$ is a *filtering operation* applied over D, which defined as follows:

$$\langle Cond \rangle = \varnothing | \{\langle B, SP \rangle\}^* \tag{3}$$

such that, B is an attribute in D and SP is a simple predicate to be applied on B.

On the other hand, Fig. 7 shows the logical model of the TLAQ query execution.

As shown in Fig. 7, the proposed logical model for TLAQ enables efficient *data exploration* and *visualization* through a *multi-step query execution flow*. Specifically, the model consists of the following stages.

- *Selection of Visualization Attributes*: The first step involves the selection of relevant attributes that will be used to visualize data. This decision shapes the scope of the subsequent query and focuses the analysis on meaningful data dimensions. *Visualization attributes* are typically chosen based on the analytic goals, providing context for the data distribution analysis;
- *Analytical Query Editing*: After identifying the visualization attributes, users engage in editing the analytical query. This allows for further specification of the *tree-like structure* of the query by selecting subqueries that filter data based on *predicates, operators,* and *attribute filter values*. This stage ensures that only relevant subsets of the data are processed, aligning with the desired analytical insights;
- *Node Query Execution*: The query execution begins with *Breadth-First Search* (BFS) *tree traversal*, where each node of the tree represents a subquery that filters and

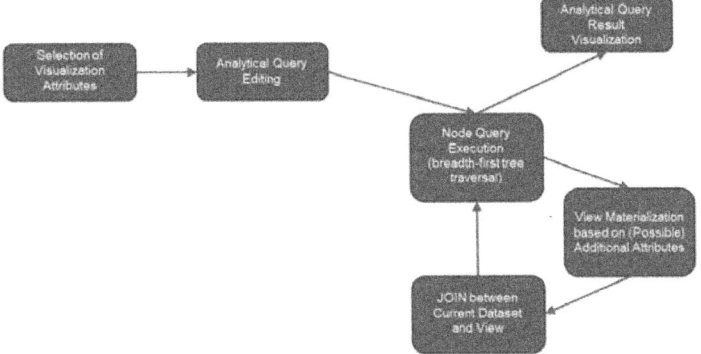

Fig. 7. Logical Model of TLAQ Query Execution.

aggregates data based on specified *conditions*. The breadth-first approach ensures that queries are executed in a *level-by-level manner*, making it easier to visualize relationships between the various nodes and their data distributions;
- *Analytical Query Result Visualization*: Once the node query execution is completed, the results are visualized according to the selected attributes. The visualization process uses techniques such as *graphs, plots*, or *tables* to display the *data distribution*, helping us identify *trends, correlations*, and *anomalies* in the dataset;
- *View Materialization based on Additional Attributes*: To further enrich the analysis, the model introduces the *materialization* of *views* that include additional attributes, which were either previously unselected or derived through further processing. These additional attributes may provide deeper insights into data trends and facilitate a more *granular* examination of the dataset;
- *JOIN between Current Dataset and View*: After materializing the view, a JOIN operation is performed between the current dataset and the newly created view. This JOIN enables the integration of additional data attributes with the existing results, which expands the dataset and provides a comprehensive *view* of the analytical query outcomes. Finally, after performing the JOIN, *breadth-first tree traversal* is re-applied to the modified dataset to refine and visualize the new results.

Specifically, Fig. 8 shows Algorithm `TLAQ Model`, which summarizes the TLAQ model.

As shown in Fig. 8, Algorithm `TLAQ Model` takes as input: (*i*) a dataset D; (*ii*) a set of dataset attributes $\mathcal{A}$; (*iii*) a query Q, and returns as output, a set of results $\mathcal{R}$.

Algorithm 1 TLAQ Model
___
Input: Dataset D, Dataset Attributes $\mathcal{A}$, Query Tree Q
Output: Set of Results $\mathcal{R}$

Begin
$\quad A_{sel} \leftarrow selectAtt(\mathcal{A})$;
$\quad Q \leftarrow editQuery()$;
$\quad Visited \leftarrow \emptyset$;
$\quad D' \leftarrow D$;
$\quad$ **for** $(v \in Q)$ **do**
$\quad\quad$ **if** $(v \notin Visited)$ **then**
$\quad\quad\quad Stack \leftarrow \emptyset$;
$\quad\quad\quad Stack.push(v)$;
$\quad\quad\quad$ **while** $(Stack \neq \emptyset)$ **do**
$\quad\quad\quad\quad u \leftarrow Stack.pop()$;
$\quad\quad\quad\quad$ **if** $(u \notin Visited)$ **then**
$\quad\quad\quad\quad\quad Visited.add(u)$;
$\quad\quad\quad\quad\quad result \leftarrow execute(u, D')$;
$\quad\quad\quad\quad\quad \mathcal{R}.add(visualize(result, A_{sel}))$;
$\quad\quad\quad\quad\quad$ **for** $(w \in Neighbors(u))$ **do**
$\quad\quad\quad\quad\quad\quad$ **if** $(w \notin Visited)$ **then**
$\quad\quad\quad\quad\quad\quad\quad Stack.push(w)$;
$\quad\quad\quad\quad\quad\quad$ **end if**
$\quad\quad\quad\quad\quad$ **end for**
$\quad\quad\quad\quad$ **end if**
$\quad\quad\quad\quad A_{add} \leftarrow selectAtt(\mathcal{A})$;
$\quad\quad\quad\quad V_{mat} \leftarrow materialize(A_{add})$;
$\quad\quad\quad\quad D' \leftarrow D' \bowtie V_{mat}$;
$\quad\quad\quad$ **end while**
$\quad\quad$ **end if**
$\quad$ **end for**
$\quad$ **return** $\mathcal{R}$;
End

Fig. 8. Algorithm TLAQ Model.

4 Tree-Like Analytical Query Model in Practice: A Case Study

In this Section, we provide a running example to demonstrate how the TLAQ model operates through the execution of the QFLS federated system, as depicted in Fig. 9, with a particular focus on the *IMMUCARE* real-life dataset. This dataset is stored in a federated manner across two nodes within the QFLS system, one located in Portugal and the other in Spain. On top of this federated dataset, we define and execute several example TLAQ queries.

More specifically, the *IMMUCARE* dataset is a comprehensive clinical and biological dataset designed to support research on immune-related diseases, particularly in the context of immunotherapy and cancer treatments. *IMMUCARE* has a main data table named IMMUCARE_INCLUSION_En, which is shown in Fig. 10.

As shown in Fig. 10, the *IMMUCARE* database stores data about patient clinical and biological data designed for research in immuno-oncology. It includes detailed patient information, such as demographic details, medical history, laboratory results, treatments, and clinical outcomes, particularly focusing on cancer patients undergoing immunotherapy. The dataset is structured to facilitate analysis of various medical conditions, therapeutic responses, and disease progression. We have extracted the anonymized version of

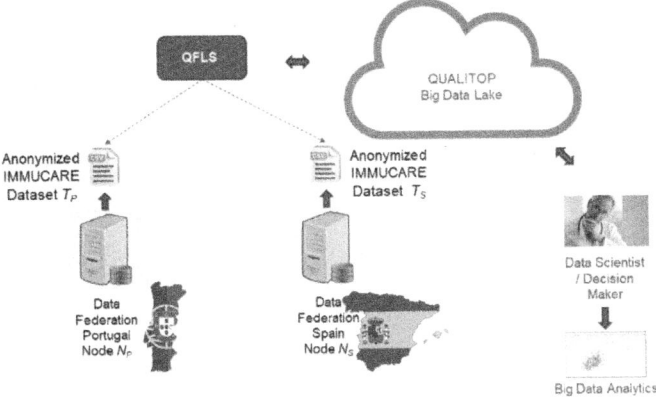

Fig. 9. The QFLS Federated System Architecture.

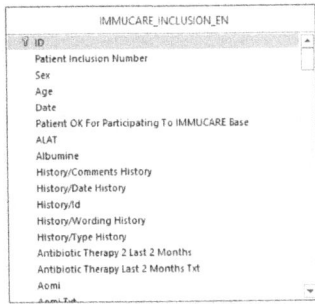

Fig. 10. The *IMMUCARE* SQL-Server Database.

the main *IMMUCARE* table (anonymized using k-anonymity, $k = 2$), which presents a subset of data from the original *IMMUCARE* table, where sensitive patient information has been anonymized to ensure k-anonymity with $k = 2$. The anonymization process ensures that each entry is indistinguishable from at least one other entry in the dataset, thus enhancing privacy while maintaining the utility of the data for analysis. The table includes anonymized attributes and other relevant clinical data, preserving the integrity of the dataset for research purposes.

In QFLS system, the first operation to set is the selection of the *visualization attributes*, i.e. the main attributes (of interest for the underlying precision medicine process) with respect to which the *advanced aggregate analytics* is computed, for every query node of the current TLAQ.

Figure 11 shows the visualization attributes for the *IMMUCARE* dataset.

☒ Date	☒ Chemo Tried Line Start Date
☒ Age	☒ Chemo Tried Line End Date
☒ Chemo Tried Line	☒ Concomitant Cancer Loc. Incl Cle
☒ Chemo Tried Line Change	☒ Concomitant Cancer Loc. Incl/Code

Fig. 11. Visualization Attributes for the *IMMUCARE* Dataset.

Figure 12 shows the first TLAQ on the *IMMUCARE* database, namely $TLAQ_{IMMUCARE,0}$.

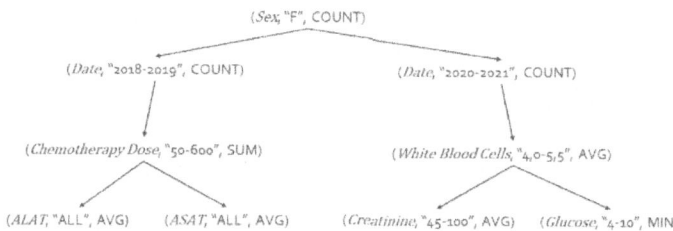

Fig. 12. The TLAQ $T_{IMMUCARE,0}$.

As shown in Fig. 12, $TLAQ_{IMMUCARE,0}$, at root node N_0, queries all the *Female* patients, and retrieves the COUNT aggregation over them. Then, at the first level, $TLAQ_{IMMUCARE,0}$, at node $N_{1.1}$, queries, within the precedent root dataset view, the date range *2018–2019* for *IMMUCARE* patient records, while, at node $N_{1.2}$, the date range *2020–2021*.

At the second level, $TLAQ_{IMMUCARE,0}$, at node $N_{2.1}$, queries, within the precedent dataset view $N_{1.1}$, patient records (records dating between *2018–2019*) that have the sum of chemotherapy dose in the range *50–600*, while, at node $N_{2.2}$, queries, within the precedent dataset view $N_{1.1}$, patient records (records dating between *2020–2021*) that have an average number of white blood cells ranging between *4.0–5.5*.

Finally, at the third level, $TLAQ_{IMMUCARE,0}$, at node $N_{3.1}$, queries, within the precedent dataset view $N_{2.1}$, patient records (having the sum of chemotherapy dose in the range *50–600* and records dating between *2018–2019*) the aggregation of average of ALAT for all these patients, while, at node $N_{3.2}$, queries, within the precedent dataset view $N_{2.1}$, patients (having the sum of chemotherapy dose in the range *50–600* and records dating between *2018–2019*) the aggregation of average of ASAT for all these patients. Still at the third level, $TLAQ_{IMMUCARE,0}$, at node $N_{3.3}$, queries, within the precedent dataset view $N_{2.2}$, patients (having an average number of white blood cells ranging between *4.0–5.5* and records dating between *2020–2021*) that have an average value of creatinine ranging between *45–100*, while, at node $N_{3.4}$, queries, within the precedent dataset view $N_{2.3}$, patients (having an average number of white blood cells ranging between *4.0–5.5* and records dating between *2020–2021*) the patient that has the lowest value of glucose in the range *4–10*.

Following that, for the TLAQ $TLAQ_{IMMUCARE,0}$, we demonstrate the complete execution of the root node query, as an example, by also providing the associated aggregate analytics. Figure 13 shows the execution at node N_0.

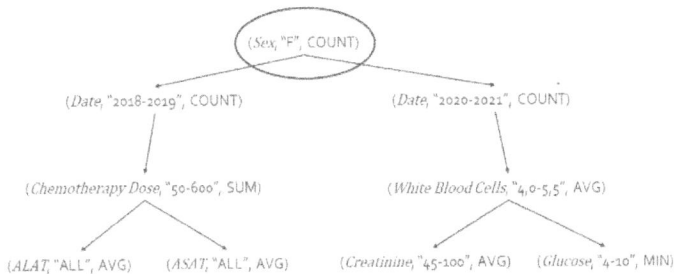

Fig. 13. TLAQ $T_{IMMUCARE,0}$: Execution at Node N_0.

Figure 14 shows the analytics related to the execution at node N_0, particularly the SQL statement for creating the big data view at node N_0. Obviously, in the QFLS system, these SQL statements are implemented via suitable *Cloud-aware Apache Spark SQL libraries* (Apache Spark).

```
CREATE VIEW Dataset-Q0(ID, Date, Age, [Tried Chemo Lines
Incl/Chemo Tried Line], [Tried Chemo Lines Incl/Chemo Line Tried Line
Change], [Tried Chemo Lines Incl/Chemo Line Tried Start Date], [Tried
Chemo Lines Incl/Chemo Line Tried End Date], [Concomitant Cancer
Localisation Incl Cle], [Concomitant Cancer Localisation Incl/Code]) AS
(SELECT ID, Date, Age, [Tried Chemo Lines Incl/Chemo Tried Line],
[Tried Chemo Lines Incl/Chemo Line Tried Line Change], [Tried Chemo
Lines Incl/Chemo Line Tried Start Date], [Tried Chemo Lines
Incl/Chemo Line Tried End Date], [Concomitant Cancer Localisation
Incl Cle], [Concomitant Cancer Localisation Incl/Code]
FROM IMMUCARE_INCLUSION_EN
WHERE Sex = "F";);
```

Fig. 14. TLAQ $T_{IMMUCARE,0}$: Big Data View SQL Statement at Node N_0.

The retrieved big data view is shown in Fig. 15.

Figure 16 shows the analytics related to the execution at node N_0, particularly the SQL statement for retrieving the aggregate value from the big data view at node N_0. The retrieved result is 332.

Similarly, we do the same analytical process for all the query nodes and we obtain the corresponding analytical results based on the particular query analysis criterion, from which in our case there are the visualization attributes.

Fig. 15. TLAQ $T_{IMMUCARE,0}$: Big Data View at Node N_0.

SELECT COUNT(*)
FROM Dataset-Q0;

Fig. 16. TLAQ $T_{IMMUCARE,0}$: Aggregate Analytics SQL Statement at Node N_0.

Figure 17 provides the final result of TLAQ $TLAQ_{IMMUCARE,0}$, where, for each query node, the aggregate analytics is shown. The QFLS system provides powerful analytics over the TLAQ model, with particular regard to advanced aggregate analytics. Next, we provide some examples of this functionality.

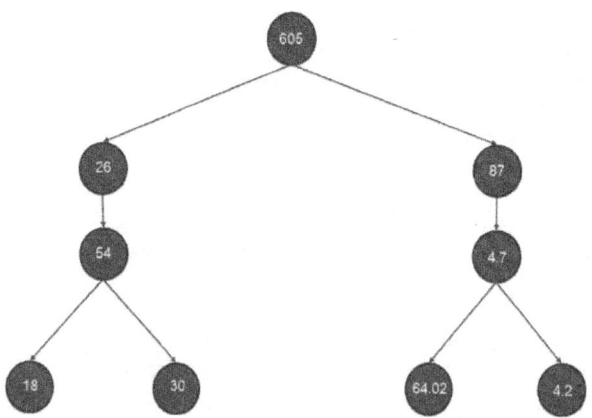

Fig. 17. TLAQ $T_{IMMUCARE,0}$: Final Result.

Figure 18 shows, for some of the TLAQ query nodes based on the various visualization attributes, the advanced aggregate analytics associated with the nodes.

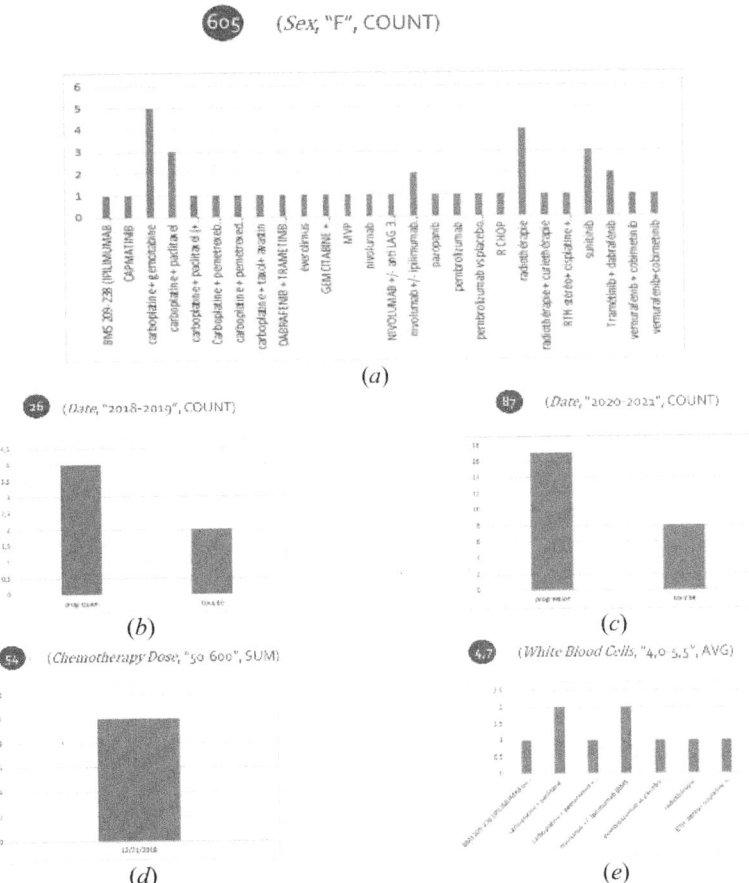

Fig. 18. TLAQ Query Advanced Aggregate Analytics Based on the Various Visualization Attributes.

5 Conclusions and Future Work

In this paper, we introduce the *Tree-Like Analytical Queries* model, which enhances big healthcare data analytics through the innovative concept of lazy aggregations. By relaxing the containment relationship between *parent-child nodes* in hierarchical *tree-like aggregate queries*, the TLAQ model enables more detailed and efficient implementations of medical investigations, particularly in epidemiological research. The case study from the *real-life EU H2020 research project* demonstrates the model's effectiveness and practical potential.

Future work is mainly oriented through exploring the applicability of the TLAQ model across different *healthcare domains*, as well as investigating *privacy* and *security* aspects of data (e.g., [23–26]).

Acknowledgments. This research is supported by the ICSC National Research Centre for High Performance Computing, Big Data and Quantum Computing within the NextGenerationEU program (Project Code: PNRR CN00000013).

References

1. Bi, H., Liu, J., Kato, N.: Deep learning-based privacy preservation and data analytics for IoT enabled healthcare. IEEE Trans. Industr. Inf. **18**(7), 4798–4807 (2022)
2. Chen, L., Yang, J.J., Wang, Q., Niu, Y.: A framework for privacy-preserving healthcare data sharing. In: 14th IEEE International Conference on e-Health Networking, Applications and Services, pp. 341–346 (2012)
3. European Committee, The EU H2020 QUALITOP Project. https://h2020qualitop.liris.cnrs.fr/wordpress/index.php/project/
4. Kosorok, M.R., Laber, E.B.: Precision medicine. Ann. Rev. Statist. Appl. **6**, 263–286 (2019)
5. Gray, J., et al.: Data cube: a relational aggregation operator generalizing group-by, cross-tab, and sub totals. Data Min. Knowl. Disc. **1**(1), 29–53 (1997)
6. Cuzzocrea, A., Soufargi, S.: QFLS: a cloud-based framework for supporting big healthcare data management and analytics from big data lakes: definitions, requirements, models and techniques. In: 12th International Conference on Data Science, Technology and Applications, pp. 422–428 (2023)
7. Cuzzocrea, A., Soufargi, S.: Supporting big healthcare data management and analytics: the cloud-based QFLS framework. In: 5th International Conference on Big Data Analytics and Knowledge Discovery. pp. 372–379 (2023)
8. Gu, Z., et al.: A systematic overview of data federation systems. Semantic Web **15**(1), 107–165 (2024)
9. Chen, Y., Pang, X., Li, X., Wang, H., Niu, B., Hu, S.: U-DPAP: utility-aware efficient range counting on privacy-preserving spatial data federation. Proc. ACM Manage. Data **3**(1), 1–25 (2025)
10. Leung, C.K., Braun, P., Cuzzocrea, A.: AI-based sensor information fusion for supporting deep supervised learning. Sensors **19**(6), 1345 (2019)
11. Howlader, P., Pal, K.K., Cuzzocrea, A., Kumar, S.D.M.: Predicting facebook-users' personality based on status and linguistic features via flexible regression analysis techniques. In: 33rd ACM Annual Symposium on Applied Computing, pp. 339–345 (2018)
12. Camara, R.C., et al: Fuzzy logic-based data analytics on predicting the effect of hurricanes on the stock market. In: 2018 IEEE International Conference on Fuzzy Systems, pp. 1–8 (2018)
13. Petrosino, L., Masi, L., D'Antoni, F., Merone, M., Vollero, L.: A zero-knowledge proof federated learning on DLT for healthcare data. J. Parallel Distribut. Comput. **196**, 104992 (2025)
14. Cuzzocrea, A.: Big data lakes: models, frameworks, and techniques. In: 2021 IEEE International Conference on Big Data and Smart Computing, pp. 1–4 (2021)
15. Ganopoulou, M., et al.: Delving into causal discovery in health-related quality of life questionnaires. Algorithms **17**(4), 138 (2024)
16. Gonçalves, H.R., Pinheiro, P.G.C.D., Pinheiro, C., Martins, L., Rodrigues, A.M., Santos, C.P.: Deep learning models for improving parkinson's disease management regarding disease stage, motor disability and quality of life. Comput. Biol. Med. **189**, 109961 (2025)
17. Cuzzocrea, A., De Bock, G.H., Maas, W.J., Soufargi, S.: F-TBDA: a frequency-based temporal big data analytics technique for mining and analyzing quality-of-life indicators of cancer patients. In: 2023 IEEE International Conference on Big Data, pp. 5197–5205 (2023)

18. Agrawal, D., Das, S., El Abbadi, A.: Big data and cloud computing: current state and future opportunities. In: 14th ACM International Conference on Extending Database Technology, pp. 530–533 (2011)
19. Ouf, S., Nasr, M.: Cloud computing: the future of big data management. Int. J. Cloud Appl. Comput. **5**(2), 53–61 (2015)
20. Sachin, D.N., Basava, A., Hegde, S., Abhijit, C.S., Ambesange, S.: FedCure: a heterogeneity-aware personalized federated learning framework for intelligent healthcare applications in IoMT environments. IEEE Access **12**, 15867–15883 (2024)
21. Wang, H., Yang, T., Ding, Y., Tang, S., Wang, Y.: VPPFL: verifiable privacy-preserving federated learning in cloud environment. IEEE Access **12**, 151998–152008 (2024)
22. Cuzzocrea, A.: Privacy and security of big data: current challenges and future research perspectives. In: 1st ACM International Workshop on Privacy and Security of Big Data, pp. 45–47 (2014)
23. Langone, R., Cuzzocrea, A., Skantzos, N.: Interpretable anomaly prediction: predicting anomalous behavior in industry 4.0 settings via regularized logistic regression tools. Data Knowl. Eng. **130**, 101850 (2020)
24. Masum, M., et al.: Bayesian hyperparameter optimization for deep neural network-based network intrusion detection. In: IEEE International Conference on Big Data, pp. 5413–5419 (2021)
25. Faruk, M.J.H., et al.: Malware detection and prevention using artificial intelligence techniques. In: 2021 IEEE International Conference on Big Data, pp. 5369–5377 (2021)
26. National Cancer Registration and Analysis Service, Health Data Insight: The Simulacrum. https://www.cancerdata.nhs.uk/simulacrum
27. Benesty, J., Chen, J., Huang, Y., Cohen, I.: Pearson correlation coefficient. In: Noise Reduction in Speech Processing, Springer (2009)
28. Sedgwick, P.: Spearman's rank correlation coefficient. BMJ **349**, 7327 (2014)

Leveraging Large Language Models for Smart Educational Services: A Comprehensive Review

Yi Li[1], Tongsong Liu[1], and Wanshou Yang[2(✉)]

[1] Shenzhen Institute of Information Technology, Longgang District, Shenzhen, China
[2] Shenzhen Polytechnic University, Nanshan District, Shenzhen, China
yangws@szpu.edu.cn

Abstract. This comprehensive review explores the integration of Large Language Models (LLMs) in smart educational services, highlighting their potential to revolutionize teaching and learning through personalized, interactive experiences. We propose a conceptual framework categorizing LLMs-Driven services into infrastructure, business processes, data management, applications, and security, privacy and ethics. Despite promising applications in adaptive tutoring, content generation, and assessment, LLMs face challenges related to accuracy, bias, pedagogical understanding, and ethical deployment. Future opportunities include enhancing model accuracy, developing pedagogically aware LLMs, and establishing robust ethical frameworks. This review aims to provide valuable insights for educators, researchers, and developers to leverage LLMs effectively and responsibly in education.

Keywords: LLMs · Smart Educational Services · Artificial Intelligence

1 Introduction

With the rapid advancement of artificial intelligence technologies, LLMs have demonstrated significant potential across various domains, with the field of smart educational services being particularly noteworthy. Education serves as a cornerstone of societal development, and the emergence of smart educational services has brought about profound transformations in traditional educational models. These services are capable of offering personalized learning experiences and enhancing teaching efficiency and quality through intelligent means, thereby catering to the diverse needs of learners.

LLMs have emerged as a pivotal force in propelling the development of smart educational services, thanks to their powerful capabilities for language understanding and generation. These models, trained on vast amounts of text data, have accumulated extensive knowledge and linguistic expression abilities, enabling them to comprehend complex natural language instructions and generate coherent and accurate textual content. Within educational contexts, LLMs can assume various roles, such as virtual teachers, learning assistants, and intelligent tutoring tools, providing students with instant answers to questions, learning suggestions, and knowledge expansion services.

However, despite the promising application prospects of LLMs in smart educational services, the related works are scattered and lack a systematic framework to examine them efficiently. Considering these points, in this comprehensive review, we propose a conceptual framework of smart educational services and conduct an indepth investigations for the existing research works based on this framework. We also examine the challenges and limitations, and future directions and opportnities of LLMs in smart educational services. By systematically summarizing and analyzing existing research findings, we hope to provide valuable references and guidance for educators, researchers, and technology developers. Our ultimate goal is to promote further development and application of LLMs in the field of smart educational services, thereby contributing to the intelligent transformation of education.

The remainder of this paper is organized as follows: Sect. 2 gives an overview of LLMs. Subsequently, a conceptual framework of smart educational services and state-of-the-art studies are specified in Sect. 3. Furthermore, we describe the challenges and limitations in Sect. 4. Section 5 gives future directions and opportunities. Finally conclusions are illustrated in Sect. 6.

2 Overview of LLMs

LLMs [20] have become a cornerstone in the field of artificial intelligence, particularly in natural language processing (NLP). These models are characterized by their massive scale, often comprising billions of parameters, which enable them to learn complex patterns and generate human-like text. LLMs have demonstrated significant capabilities in various tasks, including text generation, translation, summarization, and dialogue systems.

The key features of LLMs are summarized as follows [20]:

- Expressive Power: LLMs can capture intricate language patterns and semantics, allowing them to understand and generate coherent and contextually relevant text.
- Generalization Ability: They exhibit strong performance in both zero-shot and few-shot learning scenarios, where they can adapt to new tasks with minimal or no additional training.
- Multimodal Capabilities: Some advanced LLMs can process not only text but also other modalities such as images and audio, expanding their application scope.
- Pretraining and Fine-tuning: LLMs are typically pretrained on vast amounts of unlabeled data and then fine-tuned for specific tasks, which allows them to leverage general language knowledge while adapting to particular domains.

LLMs have been widely applied across multiple domains, as illustrated in Table 1.

Table 1. Applications of LLMs

Application Domain	Description
Healthcare [14,28]	Assisting in medical text analysis, patient education, and aiding in clinical decision-making.
Text Generation [4,42]	Creating coherent and contextually appropriate text for various purposes, such as writing articles, stories, or even code.
Translation and Summarization [8,49]	Facilitating cross-language communication and condensing large volumes of text into concise summaries.
Entertainment [10,30]	Generating scripts, dialogues, and interactive content for games and other media.
Customer Service [32,35]	Automating customer support through chatbots and virtual assistants, providing instant responses to user queries.

3 Smart Educational Services: A Conceptual Framework

Smart educational services [38] involve integration of advanced technologies, such as artificial intelligence (AI), machine learning, and data analytics, into educational systems to enhance teaching and learning experiences. These services aim to provide personalized, adaptive, and interactive learning environments that can cater to the diverse needs of students. They often include features like intelligent tutoring systems, adaptive assessments, and virtual learning environments that can dynamically adjust to the learnerâĂŹs progress and preferences.

The advent of LLMs has significantly impacted the development and capabilities of smart educational services. LLMs, with their advanced natural language processing (NLP) capabilities, can understand and generate human-like text, making them ideal for various educational applications. For instance, LLMs can serve as virtual teaching assistants, providing real-time feedback and explanations to students. They can also generate personalized learning materials, such as summaries, quizzes, and practice exercises, tailored to the individualâĂŹs learning pace and style.

Moreover, LLMs can enhance interactive learning experiences by facilitating natural language interactions between students and educational systems. This can include answering complex questions, providing explanations, and even engaging in dialogues that simulate human tutoring. Additionally, LLMs can analyze large volumes of educational data to identify patterns and insights that can inform instructional strategies and improve educational outcomes.

However, despite the promising application prospects of LLMs in smart educational services, these researches lacks unified framework, to examine the LLMs in smart education services, we present a conceptual framework to review exsiting works.

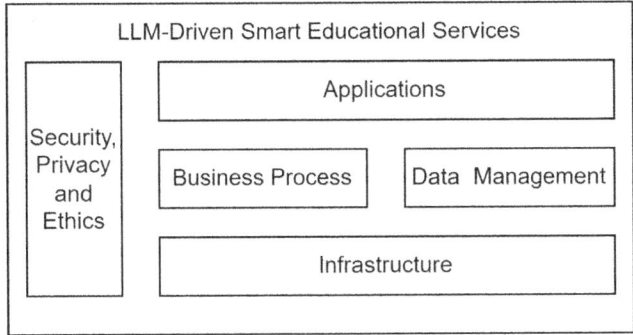

Fig. 1. A Conceptual Framework for Smart Educational Services.

3.1 The Framework

The framework of LLMs-Driven smart educational services can be divided into five key parts, which are infrastructure, business process, data management, applications and security, privacy and ethics. The infrastructure refers to technical infrastructures or platforms to enable the LLMs-Driven smart educational services. The business process is the core business flow for smart education services. The data management refers to the data storage, computing and governance in smart education services. The applications depict the specific applications of smart education services in real scenarios. The relationships of these components can be demostrated in Fig. 1.

3.2 Infrastructure

The infrastructure of LLMs-Driven smart educational services is a complex ecosystem that integrates various components to support the effective deployment and operation of LLMs in educational contexts. Below is a categorized review of the key components of this infrastructure:

Computing Infrastructure. Computing infrastructure refers to the computing devices for building the LLMs in smart educational services. Zesch (2024) [48] presents a LLM experimental infrastructure to support experimentation and innovation in higher education. They provides an open and locally available LLM server for supporting the education services. Also to address the inequality of education, some researchers try to use mobile phones to run the LLMs, which can widen the gap, students in underprivileged contexts would miss the new opportunities brought about by the use of LLMs in education [27].

Training Infrastructure. The development of LLMs has significantly impacted the education sector, with various platforms and frameworks emerging to support their training and application. Xu [44] presents an open-source platform for developing, training, and evaluating LLM-based recommender systems. It supports both encoder-decoder models and decoder-only models across 10 public datasets for sequential and straightforward recommendations. The platform includes three item indexing methods (random, sequential, and collaborative indexing) and is built on the Transformers library for easy customization. The work in [36] introduces AxoNN, a novel four-dimensional hybrid parallel algorithm in an open-source framework, optimized for matrix multiplication, computation overlap, and performance modeling. AxoNN achieves unprecedented scaling and peak flop/s for training GPT-style models on Perlmutter, Frontier, and Alps. The study explores catastrophic memorization and proposes a solution, demonstrating fine-tuning of a 405-billion parameter LLM on Frontier using AxoNN. The work in [2] introduces vTrain, a profiling-driven simulator that helps AI practitioners identify optimal parallelization strategies, balance training time and cost, and design compute-efficient LLM architectures within budget constraints.

Inferring Infrastructure. The inferring infrastructure aims to offer the serving environment for LLMs in smart education. It closely relates to the performance of LLMs applications. Some works focused on this area. The work in [6] developed an effective simulation tool, called LLMServingSim, to support future research in LLM serving systems, which can simulate the LLM serving in the granularity of iterations, leveraging the computation redundancies across decoder blocks and reusing the previous simulation results. To improving the efficiency of inferring, Yu .et al. [46] propose a new computing paradigm to accommodate the twin computing engines (GPU and CPU) and the hierarchical memory architecture (GPU and CPU memory) in an asymmetric multiprocessing framework named TwinPilots. This framework enables an effective scheduling that balances the speeds of the CPU and the GPU.

Agents. Agents [15] in the context of artificial intelligence refer to a system or software entity that perceives its environment through sensors and acts upon that environment through actuators to achieve specific goals. Agents can be designed to operate autonomously, making decisions based on their inputs and predefined objectives. They are often equipped with capabilities such as natural language understanding, decision-making, and learning from interactions.

In the field of education, agents have become increasingly important as tools to enhance teaching and learning experiences. Hu et al. (2024) [16] present an intelligent system designed to assist both teaching and learning by integrating multimodal large models and a multi-agent framework. It leverages intelligent hardware to record real classroom videos, which are then processed by multimodal models for recognition, comprehension, and description. Sena et al. [29] employs LLMs-Driven agents to assist the language learning. Each agent specializes in a specific function, working together to provide personalized, adaptive learning support. The work in [26] proposes an intelligent chatbot tutoring system that integrates Retrieval-Augmented Generation (RAG) with a custom LLM to deliver accurate, context-aware, and tailored learning support. By retrieving information from curated academic sources, incorporating interactive feedback, and using machine learning for continuous improvement, the system enhances student comprehension, engagement, and academic performance. Expected outcomes include improved personalized learning, knowledge retention, and educational outcomes, positioning the AI tutor as a key tool in modern education.

3.3 Business Process

Business process in traditional concept [39] is regarded as a collection of activities that take one or more kinds of input and create an output that is of value to the customer. The integration of LLMs into smart education services transforms traditional business processes by enabling automation, personalization, and scalability. Key stages include:

- **Content Curation and Knowledge Retrieval** [5]: LLMs, combined with Retrieval-Augmented Generation (RAG), fetch relevant educational materials from structured databases or institutional repositories to ensure accuracy and alignment with curriculum standards.
- **Personalized Learning Delivery** [31]: AI tutors analyze student queries, learning history, and performance data to provide adaptive explanations, exercises, and feedback, mimicking human tutoring while maintaining consistency.
- **Interactive Engagement and Feedback Loop** [25]: Chatbots engage students through natural language interactions, while collecting real-time feedback to refine responses and improve service quality via machine learning optimization.
- **Assessment and Performance Analytics** [21]: LLMs assist in automated grading, progress tracking, and predictive analytics, helping educators identify knowledge gaps and tailor interventions.
- **Scalability and Institutional Integration** [34]: Cloud-based deployment allows seamless integration with Learning Management Systems (LMS), enabling cost-effective, 24/7 tutoring support for large student populations.

3.4 Data Management

Data management is also a critical part for smart educational services due to the fact that the LLMs heavily rely on the data quality [11]. Educational data management involves not only the collection, storage, and analysis of data but also has a close relationship with students' learning experiences, teachers' instructional decision-making, and the formulation of educational policies. In recent years, with the rise of artificial intelligence, big data, and machine learning technologies, the role of educational data management in improving educational quality and personalized learning has become increasingly significant.

Data Collection and Storage. Data collection is a fundamental aspect of educational data management. Research indicates that educational data come from a wide range of sources, including students' learning behavior logs, interaction records from online learning platforms, and examination scores. These data are collected and integrated through various tools and technologies to build comprehensive educational data warehouses. For example, some studies [13] analyze students' learning behaviors using clickstream data from Learning Management Systems (LMS). Regarding data storage, the importance of data security and privacy protection is emphasized. With the increasing volume of data, distributed storage and cloud computing technologies are widely applied to in the storage and management of educational data.

Data Analysis and Application. Data analysis is the core component of educational data management. In recent years, machine learning and artificial intelligence technologies have been widely applied in educational data analysis. For example, by analyzing students' learning behavior data, it is possible to predict their learning outcomes and provide personalized teaching suggestions for teachers. Moreover, some studies [11] have explored how to use data visualization techniques to help students better engage in self-regulated learning. Data analysis is applied not only at the student level but also for decision-making in educational institutions. By analyzing educational data, educational administrators can gain a better understanding of the operation of the educational system and optimize the allocation of educational resources.

Technological Innovations and Tools. Technological innovations [11] are constantly emerging in the field of educational data management. For example, chatbots, as an emerging technology, have been widely applied in education. Research indicates that chatbots can provide personalized learning support for students, helping them to better complete their learning tasks. In addition, some studies have explored how to optimize the performance of chatbots using big data analysis and machine learning algorithms. In addition to chatbots, learning analytics dashboards are also important tools in educational data management. These dashboards visually present educational data to help teachers and students better understand the learning process.

3.5 Applications

Applications are built based on the infrastructure, business process and data management.The main applications of LLMs-Driven smart educational services can be summarized in Table 2.

3.6 Security, Privacy and Ethics

In this section, security, privacy and ethics of LLMs-Driven smart educational services are discussed in detail. Security, privacy and ethics are integral components of the entire business process of smart educational services.

Table 2. Applications of LLMs-Driven Smart Educational Services

Category	Application	Description
Personalized Tutoring	- Adaptive Q&A systems - Step-by-step problem solving	LLMs (e.g., ChatGPT) combined with **RAG** retrieve course-specific materials to provide context-aware explanations. Custom LLMs fine-tuned on educational datasets reduce hallucinations risks [1,40].
Automated Assessment	- Essay grading - Code evaluation	Models like **GPT-4** or open-source alternatives (e.g., Llama2) analyze student submissions for coherence, accuracy, and plagiarism. Integrated with LMS for scalable feedback [4,41].
Content Generation	- Lecture summaries - Quiz generation	LLMs synthesize textbooks/lectures into concise notes or generate practice questions aligned with learning objectives [23,43].
Language Learning	- Conversational practice - Grammar correction	Chatbots simulate dialogues for language learners, while **transformers** provide real-time grammar/style suggestions [22,50].
Administrative Support	- Scheduling assistants - FAQ automation	LLMs handle institutional queries (e.g., admissions, deadlines) via **Ollama** or cloud-based APIs, reducing staff workload [19,24].
Accessibility	- Text-to-speech - Multilingual translation	Localized LLMs (e.g., **Llama.cpp**) offer low-latency translations or audio outputs for visually impaired students [7,9].

Security

- Data Protection: LLMs-Driven smart educational services handle vast amounts of sensitive data, including personal information, learning records, and academic performance. Ensuring robust encryption and secure storage mechanisms is crucial to protect this data from unauthorized access and breaches [33].
- Authentication and Authorization: Implementing strong authentication protocols and role-based access controls ensures that only authorized users (e.g., students, teachers, administrators) can access specific data and functionalities within the educational platform.
- System Integrity: Regular security audits, vulnerability assessments, and updates are essential to maintain the integrity of the system. This helps in identifying and mitigating potential security threats that could compromise the functionality and reliability of LLMs-Driven services.

Privacy

- Data Minimization: Collecting only the data that is necessary for the educational service and avoiding unnecessary data collection helps in minimizing privacy risks [12]. This ensures that the amount of personal information exposed is limited.
- Anonymization and Pseudonymization: Techniques such as anonymization and pseudonymization facilitate protect the identity of students and teachers. This helps maintain privacy while still allowing the use of data for analysis and improvement of educational services.
- Consent and Transparency: Obtaining explicit consent from users for data collection and usage is a fundamental aspect of privacy protection. Providing transparent information about how data will be used, stored, and shared ensures that users are aware of the implications and can make informed decisions.

Ethics The issues of ethics can be summarized as follows [45]:

- Bias and Fairness: LLMs are trained on large datasets that may contain biases. Ensuring that the models do not perpetuate or amplify these biases is crucial for maintaining fairness in educational services. Regular monitoring and evaluation of model outputs for potential biases are necessary to address and mitigate such issues.
- Transparency and Explainability: The decision-making processes of LLMs should be transparent and explainable, especially in educational contexts where students and teachers need to understand the basis for recommendations or assessments. This helps in building trust and ensures that the services are used appropriately.
- Responsible Use: Educators and developers should be aware of the ethical implications of using LLMs in education. This includes avoiding misuse of the technology, such as using it for surveillance or inappropriate monitoring of students. Promoting responsible use and adhering to ethical guidelines is essential for the sustainable and ethical deployment of LLMs in educational services.

4 Challenges and Limitations

Despite the transformative potential of LLMs in creating smart educational services, their deployment is fraught with significant challenges and inherent limitations that require careful consideration and mitigation strategies. These issues span technical, pedagogical, ethical, and practical domains.

4.1 Accuracy, Reliability, and Hallucinations

One of the most critical challenges is the propensity of LLMs to 'hallucinate' âĂŞ generating plausible-sounding but factually inaccurate or nonsensical information [17]. In an educational context, where accuracy is paramount, disseminating misinformation can severely undermine learning outcomes and trust. Ensuring the factual correctness and reliability of LLM-generated content, explanations, and feedback remains a major hurdle, often requiring sophisticated verification mechanisms or human oversight.

4.2 Bias and Fairness

LLMs are trained on vast datasets, predominantly sourced from the internet, which often contain inherent societal biases related to gender, race, culture, and socioeconomic status [3]. These biases can be reflected and even amplified in the educational services they power, potentially leading to stereotyped content, inequitable assessment, or unfair personalization that disadvantages certain student groups. Addressing and mitigating algorithmic bias to ensure fairness and equity in LLMs-Driven education is a complex and ongoing challenge.

4.3 Lack of True Pedagogical Understanding

While LLMs can mimic human-like conversation and generate relevant content, they lack genuine pedagogical grounding, common-sense reasoning, and a deep understanding of learning theories or individual student cognitive states [18]. They may struggle to accurately diagnose complex learning difficulties, provide truly adaptive scaffolding within a student's Zone of Proximal Development (ZPD), or exhibit the empathy and motivational skills crucial for effective teaching. Over-reliance on LLMs might lead to superficial learning or neglect the development of deeper conceptual understanding.

4.4 Cost, Accessibility, and Infrastructure

Developing, training, and deploying state-of-the-art LLMs requires substantial computational resources, specialized hardware (e.g., GPUs), and significant financial investment [37]. The operational costs for inference can also be high. This raises concerns about accessibility and equity, potentially widening the digital divide between well-resourced institutions and those with limited budgets or technical infrastructure. Integrating LLMs seamlessly into existing educational platforms also presents technical and logistical barriers.

4.5 Over-Reliance and Impact on Critical Skills

There is a potential risk that excessive reliance on LLMs for tasks like writing, problem-solving, or information retrieval could hinder the development of students' essential skills, such as critical thinking, creativity, independent research, and information literacy [47]. Students might become passive recipients of information rather than active learners, potentially leading to 'deskilling'. Striking a balance between leveraging LLMs as supportive tools and fostering core cognitive abilities is crucial.

4.6 Evaluation and Assessment Limitations

While LLMs show promise in automated grading for certain task types, evaluating complex, creative, or nuanced student work (e.g., essays requiring deep analysis, novel problem-solving approaches) remains challenging. LLMs may struggle to accurately appreciate subtlety, originality, or diverse perspectives, potentially leading to standardized or superficial assessments. Ensuring that LLM-based assessments are valid, reliable, and fair requires careful design and validation.

In conclusion, while LLMs offer exciting avenues for innovation in educational services, overcoming these multifaceted challenges is essential for their responsible, effective, and equitable implementation. Future research and development must prioritize not only technological advancement but also robust solutions addressing accuracy, bias, pedagogy, privacy, cost, and the holistic development of learners.

5 Future Directions and Opportunities

While the challenges associated with LLMs in education are substantial, the potential opportunities for innovation and transformation are equally compelling. Addressing the current limitations opens up numerous exciting avenues for future research and development, paving the way for more effective, equitable, and engaging smart educational services.

5.1 Enhancing Accuracy, Reasoning, and Controllability

Future research will likely focus on developing techniques to improve the factual accuracy and logical reasoning capabilities of LLMs. These techniques are summarized as:

- Neuro-Symbolic Approaches: Combining the pattern-recognition strengths of neural networks with the logical reasoning capabilities of symbolic AI systems in order to create hybrid models with enhanced reliability.
- Improved Controllability: Developing finer-grained controls over LLM outputs, allowing educators or designers to specify constraints regarding tone, complexity, pedagogical strategy, or adherence to specific curricula, thus mitigating risks of inappropriate or inaccurate content generation.

5.2 Developing Pedagogically-Aware LLMs

A key opportunity lies in imbuing LLMs with a deeper understanding of pedagogical principles and learning science. This includes three parts:

- Student Cognitive Modeling: Training LLMs to infer student knowledge states, misconceptions, engagement levels, and affective states more accurately based on interaction data, enabling more precise adaptation.
- Explicit Pedagogical Strategy Implementation: Designing LLMs capable of consciously selecting and implementing specific teaching strategies (e.g., Socratic questioning, worked examples, inquiry-based learning) based on the learning objectives and the inferred needs of the student.
- Affective Computing Integration: Incorporating capabilities to recognize and appropriately respond to student emotions (e.g., frustration, curiosity, boredom) to provide socio-emotional support and maintain motivation.

5.3 Towards True Hyper-Personalization at Scale

LLMs offer the potential to move beyond coarse-grained adaptation towards truly individualized learning pathways at an unprecedented scale. Future directions can be expressed as follows:

- Dynamic Content Generation and Sequencing: Automatically generating bespoke learning materials (texts, exercises, explanations) tailored to individual prior knowledge, learning pace, and preferred modalities (including multimodal generation).
- Adaptive Scaffolding and Feedback: Providing real-time, context-aware scaffolding that dynamically adjusts based on student performance, offering hints or simplified explanations when needed and fading support as competence grows. Feedback can become more granular, conceptual, and immediately actionable.
- Personalized Learning Trajectories: Utilizing LLMs to help students navigate complex knowledge domains, suggesting optimal learning paths based on their long-term goals and identified knowledge gaps.

5.4 Multimodal and Interactive Educational Experiences

Future LLMs are becoming increasingly multimodal, capable of processing and generating information across text, images, audio, and video. This opens opportunities for:

- Rich Interactive Simulations: Creating immersive learning environments where students can interact with LLM-powered agents or systems in realistic scenarios (e.g., virtual science labs, historical simulations, language practice conversations).

– Automated Analysis of Multimodal Work: Developing tools that can analyze and provide feedback on student work involving diagrams, presentations, code visualizations, or even physical demonstrations captured on video.
  – Accessible Learning Tools: Generating alternative representations of information (e.g., textual descriptions of images for visually impaired students, sign language interpretations of text) to enhance accessibility.

5.5 Synergistic Human-AI Collaboration for Educators

Rather than replacing teachers, a significant opportunity lies in developing LLMs as powerful assistants and collaborators for educators. This involves:

  – Teacher Co-Pilots: Creating tools that assist teachers with time-consuming tasks like lesson planning, differentiating instruction, generating assessment items, drafting feedback, and identifying students needing extra support.
  – AI-Augmented Professional Development: Providing personalized professional development resources and coaching for teachers, driven by LLM analysis of classroom practices (where ethically appropriate and consented) or self-reflection prompts.
  – Enhanced Classroom Analytics: Developing teacher dashboards that provide actionable insights into student progress, common misconceptions, and engagement patterns, powered by LLM analysis of learning data.

5.6 Robust Ethical Frameworks, Governance, and Explainability

Significant effort must be directed towards establishing robust ethical guidelines and technical solutions for responsible LLM deployment in education. It involves three parts:

  – Bias Detection and Mitigation: Developing advanced techniques to proactively identify and mitigate biases in training data and model outputs.
  – Privacy-Preserving Techniques: Exploring methods like federated learning, differential privacy, and secure enclaves to enable LLM training and use without compromising sensitive student data (Kairouz et al., 2021).
  – Explainable AI for Education: Creating LLM systems whose reasoning and decision-making processes (e.g., why a particular feedback was given, how a student model was inferred) are transparent and interpretable for educators, students, and administrators.

5.7 Exploring Novel Educational Paradigms

LLMs could enable entirely new ways of learning and assessment. The topics can be:

- Lifelong Learning Companions: Personalized AI tutors that accompany learners throughout their lives, supporting formal education, workplace training, and informal learning.
- Collaborative Problem Solving with AI: Designing learning activities where students collaborate with LLM agents to solve complex problems, fostering both domain knowledge and collaboration skills.
- Creativity Augmentation: Using LLMs as brainstorming partners or tools to help students explore creative possibilities in writing, art, music, and design.

The trajectory of LLMs in education points towards a future where learning can become significantly more personalized, interactive, efficient, and accessible. Capitalizing on these opportunities requires a concerted, interdisciplinary effort involving researchers, educators, policymakers, and developers, guided by a strong commitment to pedagogical effectiveness, ethical principles, and equitable outcomes for all learners.

6 Conclusions

This review examines potential studies of LLMs in smart educational services using a unified framework of LLMs-Driven smart educational services. Challenges such as accuracy, bias, pedagogical limitations, and ethical concerns are discussed in detail. Future directions are clearly specified include enhancing model controllability, integrating multimodal capabilities, and fostering human-AI collaboration. By advancing robust frameworks and equitable access, LLMs can revolutionize education while ensuring responsible and inclusive deployment. Interdisciplinary efforts are essential to harness their full potential for improving learning outcomes globally.

Acknowledgements. This research was supported by Research on the Innovation of Teaching Paradigm and Practical Exploration of Ideological and Political Theory Courses in Institutes Based on Large Language Model (No. 2023GXSZ169) and Research on the Improvement Strategy of Teaching Effects of Ideological and Political Theory Courses in Institutes Based on Large Language Model (No. szjy23012).

References

1. Al-Abri, A.: Exploring chatgpt as a virtual tutor: a multi-dimensional analysis of large language models in academic support. Educ. Inform. Technol. 1–36 (2025)
2. Bang, J., Choi, Y., Kim, M., Kim, Y., Rhu, M.: Vtrain: a simulation framework for evaluating cost-effective and compute-optimal large language model training. In: 2024 57th IEEE/ACM International Symposium on Microarchitecture (MICRO), pp. 153–167. IEEE (2024)
3. Bender, E.M., Gebru, T., McMillan-Major, A., Shmitchell, S.: On the dangers of stochastic parrots: can language models be too big? In: Proceedings of the 2021 ACM Conference on Fairness, Accountability, and Transparency, pp. 610–623 (2021)

4. Bevilacqua, M., et al.: When automated assessment meets automated content generation: examining text quality in the era of gpts. ACM Trans. Inf. Syst. **43**(2), 1–36 (2025)
5. Chaturvedi, N.: LLMS and NLP for generalized learning in AI-enhanced educational videos and powering curated videos with generative intelligence. In: Proceedings of the 1st Workshop on NLP for Science (NLP4Science), pp. 148–154 (2024)
6. Cho, J., Kim, M., Choi, H., Heo, G., Park, J.: Llmservingsim: a HW/SW co-simulation infrastructure for LLM inference serving at scale. In: 2024 IEEE International Symposium on Workload Characterization (IISWC), pp. 15–29. IEEE (2024)
7. Chollampatt, S., Pham, M.Q., Indurthi, S.R., Turchi, M.: Cross-lingual evaluation of multilingual text generation. In: Proceedings of the 31st International Conference on Computational Linguistics, pp. 7766–7777 (2025)
8. Donthi, S., et al.: Improving LLM abilities in idiomatic translation. In: Future of Information and Communication Conference, pp. 361–375. Springer (2025)
9. Du, M., Liu, C., Lai, J.: Instantspeech: instant synchronous text-to-speech synthesis for LLM-driven voice chatbots. In: ICASSP 2025-2025 IEEE International Conference on Acoustics, Speech and Signal Processing (ICASSP), pp. 1–5. IEEE (2025)
10. Gallotta, R., et al.: Large language models and games: a survey and roadmap. IEEE Trans. Games (2024)
11. García-López, I.M., González, C.S.G., Ramírez-Montoya, M.S., Molina-Espinosa, J.M.: Challenges of implementing chatgpt on education: systematic literature review. Int. J. Educ. Res. Open **8**, 100401 (2025)
12. Greco, D., Chianese, L.: Exploiting llms for e-learning: a cybersecurity perspective on AI-generated tools in education. In: 2024 IEEE International Workshop on Technologies for Defense and Security (TechDefense), pp. 237–242. IEEE (2024)
13. Guan, R., Raković, M., Chen, G., Gašević, D.: How educational chatbots support self-regulated learning? a systematic review of the literature. Educ. Inf. Technol. 1–26 (2024)
14. Haltaufderheide, J., Ranisch, R.: The ethics of chatgpt in medicine and healthcare: a systematic review on large language models (llms). NPJ Digit. Med. **7**(1), 183 (2024)
15. He, J., Treude, C., Lo, D.: Llm-based multi-agent systems for software engineering: literature review, vision and the road ahead. ACM Trans. Softw. Eng. Methodol. (2024)
16. Hu, Y., et al.: Memos: multimodal educational mentor and optimisation system based on multi-agent. In: 2024 Artificial Intelligence x Humanities, Education, and Art (AIxHEART), pp. 58–63 (2024)
17. Ji, Z., et al.: Survey of hallucination in natural language generation. ACM Comput. Surv. **55**(12), 1–38 (2023)
18. Kasneci, E., et al.: Chatgpt for good? on opportunities and challenges of large language models for education. Learn. Individ. Differ. **103**, 102274 (2023)
19. Kazemitabaar, M., et al.: Codeaid: evaluating a classroom deployment of an llm-based programming assistant that balances student and educator needs. In: Proceedings of the 2024 CHI Conference on Human Factors in Computing Systems, pp. 1–20 (2024)
20. Kumar, P.: Large language models (llms): survey, technical frameworks, and future challenges. Artif. Intell. Rev. **57**(10), 260 (2024)

21. Li, G., Tang, C., Chen, L., Deguchi, D., Yamashita, T., Shimada, A.: Llm-driven ontology learning to augment student performance analysis in higher education. In: International Conference on Knowledge Science, Engineering and Management, pp. 57–68. Springer (2024)
22. Li, K., Lun, L., Hu, P.: Exploring student perceptions of language learning affordances of large language models: Aq methodology study. Educ. Inf. Technol. 1–21 (2025)
23. Lohr, D., Berges, M., Chugh, A., Kohlhase, M., Müller, D.: Leveraging large language models to generate course-specific semantically annotated learning objects. J. Comput. Assist. Learn. **41**(1), e13101 (2025)
24. Lundström, O., Maleki, N., Ahlgren, F.: Online course improvement through gpt-4: Monitoring student engagement and dynamic faq generation. In: 2024 IEEE Global Engineering Education Conference (EDUCON), pp. 1–6. IEEE (2024)
25. Meyer, J., et al.: Using llms to bring evidence-based feedback into the classroom: ai-generated feedback increases secondary students text revision, motivation, and positive emotions. Comput. Educ. Artifi. Intell. **6**, 100199 (2024)
26. Modran, H.A., Bogdan, I.C., Ursu?iu, D., Samoilă, C., Modran, P.L.: Llm intelligent agent tutoring in higher education courses using a rag approach. In: International Conference on Interactive Collaborative Learning, pp. 589–599. Springer (2024)
27. Monteiro Santos, M., et al.: Near feasibility, distant practicality: empirical analysis of deploying and using llms on resource-constrained smartphones. In: Proceedings of the 13th International Conference on Information & Communication Technologies and Development, pp. 224–235 (2024)
28. Nazi, Z.A., Peng, W.: Large language models in healthcare and medical domain: a review. In: Informatics. vol. 11, p. 57. MDPI (2024)
29. Nouzri, S., EL Fatimi, M., Guerin, T., Othmane, M., Najjar, A.: Beyond chatbots: enhancing luxembourgish language learning through multi-agent systems and large language model. In: International Conference on Principles and Practice of Multi-Agent Systems, pp. 385–401. Springer (2024)
30. Papadimitriou, S., Virvou, M.: Computer games for entertainment and education: a literature review and exploration on artificial intelligence integration. Artificial Intelligence Based Games as Novel Holistic Educational Environments to Teach 21st Century Skills, 25–62 (2025)
31. Park, M., Kim, S., Lee, S., Kwon, S., Kim, K.: Empowering personalized learning through a conversation-based tutoring system with student modeling. In: Extended Abstracts of the CHI Conference on Human Factors in Computing Systems, pp. 1–10 (2024)
32. Peddinti, S.R., Katragadda, S.R., Pandey, B.K., Tanikonda, A.: Utilizing large language models for advanced service management: potential applications and operational challenges. J. Sci. Technol. **4**(2) (2023)
33. Rahman, M.A., Alqahtani, L., Albooq, A., Ainousah, A.: A survey on security and privacy of large multimodal deep learning models: teaching and learning perspective. In: 2024 21st Learning and Technology Conference (L&T), pp. 13–18. IEEE (2024)
34. Santos, E., Trigo, A.: Digital transformation in managing outgoing student applications: enhancing administrative efficiency in higher education institutions. Int. J. Bus. Process. Integr. Manag. **12**(1), 78–88 (2025)
35. Shareef, F.: Enhancing conversational ai with llms for customer support automation. In: 2024 2nd International Conference on Self Sustainable Artificial Intelligence Systems (ICSSAS), pp. 239–244. IEEE (2024)

36. Singh, S., et al.: Democratizing AI: Open-source scalable LLM training on gpu-based supercomputers. In: SC24: International Conference for High Performance Computing, Networking, Storage and Analysis, pp. 1–14. IEEE (2024)
37. Strubell, E., Ganesh, A., McCallum, A.: Energy and policy considerations for modern deep learning research. In: Proceedings of the AAAI Conference on Artificial Intelligence. vol. 34, pp. 13693–13696 (2020)
38. Tantatsanawong, P., Kawtrakul, A., Lertwipatrakul, W.: Enabling future education with smart services. In: 2011 Annual SRII Global Conference, pp. 550–556. IEEE (2011)
39. Tsakalidis, G., Vergidis, K.: Business process management: from a critical review to a contemporary ontological entity. Int. J. Bus. Process. Integr. Manag. **10**(2), 148–161 (2020)
40. Venugopalan, D., Yan, Z., Borchers, C., Lin, J., Aleven, V.: Combining large language models with tutoring system intelligence: a case study in caregiver homework support. In: Proceedings of the 15th International Learning Analytics and Knowledge Conference, pp. 373–383 (2025)
41. Wu, X., Saraf, P.P., Lee, G., Latif, E., Liu, N., Zhai, X.: Unveiling scoring processes: dissecting the differences between LLMS and human graders in automatic scoring, pp. 1–16. Technology, Knowledge and Learning (2025)
42. Wu, Y.: Large language model and text generation. In: Natural Language Processing in Biomedicine: A Practical Guide, pp. 265–297. Springer (2024)
43. Xie, T., Kuang, Y., Tang, Y., Liao, J., Yang, Y.: Using llm-supported lecture summarization system to improve knowledge recall and student satisfaction. Expert Syst. Appl. **269**, 126371 (2025)
44. Xu, S., Hua, W., Zhang, Y.: Openp5: an open-source platform for developing, training, and evaluating llm-based recommender systems. In: Proceedings of the 47th International ACM SIGIR Conference on Research and Development in Information Retrieval, pp. 386–394 (2024)
45. Yan, L., et al.: Practical and ethical challenges of large language models in education: a systematic scoping review. Br. J. Edu. Technol. **55**(1), 90–112 (2024)
46. Yu, C., Wang, T., Shao, Z., Zhu, L., Zhou, X., Jiang, S.: Twinpilots: a new computing paradigm for GPU-CPU parallel LLM inference. In: Proceedings of the 17th ACM International Systems and Storage Conference, pp. 91–103 (2024)
47. Zawacki-Richter, O., Marín, V.I., Bond, M., Gouverneur, F.: Systematic review of research on artificial intelligence applications in higher education-where are the educators? Int. J. Educ. Technol. High. Educ. **16**(1), 1–27 (2019)
48. Zesch, T., Hanses, M., Seidel, N., Aggarwal, P., Veiel, D., De Witt, C.: Flexible LLM experimental infrastructure (flexi)–enabling experimentation and innovation in higher education through access to open llms. In: 2024 21st International Conference on Information Technology Based Higher Education and Training (ITHET), pp. 1–8. IEEE (2024)
49. Zhao, H., et al.: From handcrafted features to llms: A brief survey for machine translation quality estimation. In: 2024 International Joint Conference on Neural Networks (IJCNN), pp. 1–10. IEEE (2024)
50. Zhou, H., Xu, K., Bao, Q., Lou, Y., Qian, W.: Application of conversational intelligent reporting system based on artificial intelligence and large language models. J. Theory Practice Eng. Sci. **4**(03), 176–182 (2024)

Three-Dimensional Analysis of AI-Enhanced Ideological and Political Education Instruction Within the Framework of Digital Education

Jing Chang[✉], Ying He, and Wen Tang

School of Marxism, Shenzhen Institute of Information Technology, Shenzhen 518000, Guangdong, China
15044317671@163.com

Abstract. This article examines the empowerment of ideological and political course teaching by AI within the context of educational digitalization, analyzing it from three dimensions: value implications, practical challenges, and optimization paths. At the level of value implications, AI-empowered ideological and political course teaching demonstrates innovative potential through enriched content, expanded domains, and enhanced efficiency. It can extend the boundaries of knowledge, transcend temporal and spatial constraints, and transform teaching paradigms. However, in practical implementation, the integration of AI into ideological and political course teaching encounters numerous challenges. To enhance the quality and efficiency of AI-empowered ideological and political course teaching, this paper proposes specific optimization strategies aimed at fostering the deep integration of AI with ideological and political education. These strategies seek to improve the relevance and effectiveness of ideological and political education while providing theoretical and practical support for nurturing talents capable of shouldering the significant responsibility of national rejuvenation.

Keywords: Educational Digital Transformation · Ideological and Political Education · Value Implications · Practical Challenges · Optimization Pathways

1 Introduction

In the context of global educational reform, the profound integration of digital technology is fundamentally transforming the traditional educational ecosystem. The digital revolution, with artificial intelligence as its central engine, has shifted the teaching paradigm from a one-way indoctrination model to a dynamic interaction facilitated by technological tools such as adaptive learning systems and intelligent assessment instruments. This transformation represents not only an evolution in technology but also a strategic response to the demands for talent development in the knowledge economy era. Particularly during the COVID-19 pandemic, widespread implementation of online education has expedited advancements in educational digital infrastructure. Consequently, integrated virtual and real-world teaching methods have evolved from being supplementary options to becoming essential components of education delivery. In this evolving landscape, AI transcends its role as merely a tool; it emerges as a pivotal factor driving innovation in educational concepts and restructuring models. This shift offers new opportunities to address long-standing challenges such as disparities in educational resources and inefficiencies in teaching practices.

As a pivotal course for implementing the fundamental task of moral cultivation, ideological and political education plays a central role in shaping young people's values and establishing a robust ideological foundation. The effectiveness of its teaching is intrinsically linked to the essential questions of "what kind of individuals we aim to cultivate, how we should train them, and for whom this training is intended." Traditional approaches to ideological and political instruction have long grappled with challenges such as content abstraction, oversimplified scenarios, and delayed feedback. However, advancements in AI technology—encompassing multimodal interaction, real-time data analysis, and virtual scenario construction—present a historic opportunity to overcome these obstacles. By aggregating intelligent resources to achieve theoretical concretization, leveraging algorithmic recommendations for personalized guidance, and utilizing virtual simulations to enhance emotional engagement, AI empowerment holds the potential to elevate ideological and political education from mere "knowledge transfer" to profound "value internalization." This transformation is expected not only to significantly enrich the contemporary relevance and appeal of the curriculum but also represents more than just superficial technological innovation; it constitutes an essential pathway toward realizing the substantive development of ideological and political education.

At present, the integration of artificial intelligence (AI) into ideological and political courses remains in the initial phase characterized by "technology adaptation education." On one hand, certain teaching practices tend to be mere transplants; for instance, equating intelligent tools with electronic question banks or virtual teaching aids does not address the profound transformation required in teaching methodologies. On the other hand, technical ethical risks such as algorithmic bias, data privacy concerns, and a lack of emotional engagement are gradually surfacing. More importantly, there exists an inherent tension between the instrumental rationality of intelligent systems and the humanistic value orientation intrinsic to ideological and political education. If this tension is not managed appropriately, it may result in a contradiction between "technological efficiency" and "educational failure." Therefore, it is essential to clarify the value mechanisms underpinning AI-enabled ideological and political courses while diagnosing structural contradictions within practice. Subsequently, we must explore pathways for technical integration that align with educational principles. In response to this pressing need, this study offers forward-looking and actionable solutions for AI-empowered ideological and political courses through a comprehensive three-dimensional review.

2 Related Works and Method

In the context of the rapid development of artificial intelligence technology, the innovation and reform of ideological and political teaching in colleges and universities have become an important topic of concern in the academic community. The current research focuses on the value, risks and response paths of artificial intelligence (AI) empowered ideological and political courses in colleges and universities. From the perspective of empowerment value, AI can promote the innovation of ideological and political courses from the three dimensions of "teaching", "learning" and "evaluation". Wang and Shen (2025) demonstrate that, at the teaching level, it is possible to aggregate instructional content and enhance the efficiency of teachers' lesson preparation. For instance, by systematically integrating resources, educators can construct a more coherent knowledge system [1]; Wang and Liu (2025) demonstrate that, at the learning level, it is capable of capturing students' individual needs and generating customized content. This can be achieved by utilizing algorithms to analyze learning conditions, thereby facilitating precise teaching [2]; Yu and Yang (2025) demonstrate that, at the evaluation level, the accuracy and effectiveness of teaching assessments can be enhanced through data tracking. This includes optimizing the formative evaluation of ideological and political courses by leveraging visual analysis technology [3]. However, technology empowerment comes with multiple risks. First, Liu (2025) demonstrates that the "intelligent dependence" between subjects and objects may result in a diminished educational significance. Furthermore, an excessive reliance on technology by both teachers and students can undermine emotional communication and the transmission of values among these subjects [4]; Second, Xiao (2025) demonstrates that "knowledge falsification" can lead to content distortion and pose ideological risks. Furthermore, the information disorder associated with generative AI may propagate erroneous views and influence mainstream value orientations [5]. Third, Liu and Chen (2025) demonstrate that "data segmentation" raises concerns regarding privacy ethics and information security. Furthermore, the collection and utilization of academic data may infringe upon students' rights in the absence of standardization [6]; Fourth, Xing (2025) posits that algorithmic bias and instrumental rationality may undermine the teacher-student relationship and erode social consensus. For instance, the human-like responses generated by intelligent tools could obscure the essential guiding role of educators [7]. In response to the above challenges, the academic community has proposed a systematic response: Zhou (2025) suggests that, at the level of subject literacy, it is essential to redefine the concept of "human-machine progress" in order to enhance teachers' proficiency in utilizing intelligent technology and to foster students' critical thinking skills [8]; Zhao et al. (2025) emphasize the necessity of developing a comprehensive vertical model for ideology and politics at the content construction level. It is essential to establish a corpus that aligns with ideological requirements, while also ensuring the authority and direction of teaching resources are maintained [9]. Liu, Chen, and Xing (2025) emphasize the importance of optimizing data governance by enhancing data collection practices, usage protocols, and security standards. This approach seeks to strike a balance between leveraging technical advantages and mitigating ethical risks [10]. Xue (2025) emphasizes the importance of practice modes in educational settings, advocating for a balanced integration of virtual and real environments. This approach aims to prevent an over-reliance on simulated contexts at the expense of practical education [11].

In general, existing research has thoroughly elucidated the dual effects of artificial intelligence in enhancing ideological and political courses within colleges and universities. It has also provided a practical pathway through the dimensions of value guidance, technical regulation, and subject collaboration. Building upon the theoretical findings of previous scholars, this paper will concentrate on three levels: the value implications of AI-enabled ideological and political teaching, practical dilemmas encountered, and implementation pathways. An analytical framework termed "value-challenge-path" will be constructed. This framework is not linear; rather, it embodies a spiraling dialectical relationship. The application of technological advantages may give rise to new challenges (such as algorithmic recommendations leading to information cocooning), while these challenges necessitate innovative pathways (for instance, establishing a hierarchical data permission system). Ultimately, this process aims to achieve a deep integration of AI with ideological and political education through continuous optimization. The proposed framework not only offers methodological support for the digital transformation of education but also ensures that technological empowerment consistently aligns with the fundamental goal of "cultivating individuals with virtue".

3 Value Implications

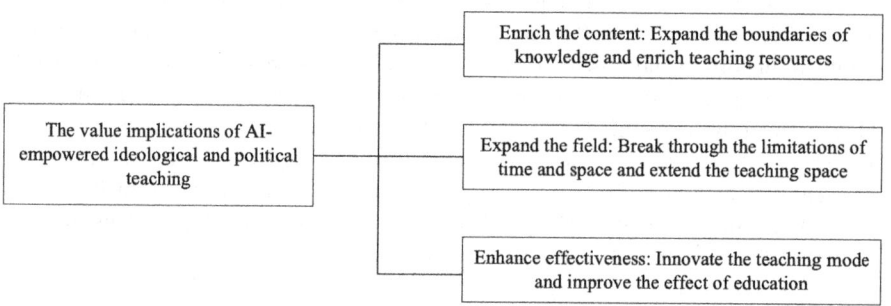

Fig. 1. The value implications of AI-empowered ideological and political teaching

3.1 Content Enrichment

The integration of artificial intelligence technology has opened new avenues for optimizing teaching resources in ideological and political theory courses. By leveraging advanced data collection and analytical capabilities, researchers can enable AI systems to swiftly aggregate global educational resources that encompass policy developments, academic research, and contemporary social issues. Through intelligent categorization, these systems create structured knowledge repositories. The extensive application of AI technology not only transcends the limitations of traditional textbooks but also broadens students' learning horizons and enriches their educational experiences. This technological advancement fosters the development of multidimensional cognitive frameworks among students. Taking the teaching practice of "Theoretical System of Socialism with Chinese Characteristics" as an example, the AI intelligent system can integrate the latest policy documents, authoritative academic perspectives, and exemplary cases to transform a previously static and limited array of teaching materials into a dynamic and diverse knowledge system. This enables students to engage with more comprehensive and varied knowledge reserves, aiding them in understanding ideological and political theories from multiple viewpoints while breaking down barriers to knowledge acquisition. Furthermore, by utilizing natural language processing technology, artificial intelligence can generate personalized teaching materials tailored to specific instructional needs. It transforms abstract theoretical concepts into concrete content that aligns with learners' interests while enhancing its appeal. This approach strengthens both the attractiveness and readability of instructional material, addresses diverse pedagogical requirements effectively, and drives continuous improvement in the quality of ideological and political education (Fig. 1).

3.2 Expand the Field

Traditional ideological and political education is often limited by fixed classroom schedules and physical spaces, resulting in subpar teaching quality and a diminished learning experience for students. However, the digital transformation of education, particularly through the extensive integration of AI technology, can effectively address these challenges and expand the temporal and spatial boundaries of ideological and political instruction. In terms of time, AI-powered online learning platforms and intelligent educational tools empower students to engage in learning at their convenience—anytime and anywhere. This flexibility allows them to transcend the constraints of traditional classroom hours while integrating ideological and political studies into their daily lives. Students can watch instructional videos or participate in online interactive discussions according to their individual learning pace and schedule. Regarding spatial dimensions, AI technology liberates education from physical confines by utilizing virtual reality (VR), augmented reality (AR), and other innovative technologies to create a hybrid teaching environment that merges virtual experiences with real-world contexts. This enables students to "immerse" themselves in authentic historical events without needing to visit historical sites physically. Furthermore, they can engage in "cross-time-and-space dialogues" with historical figures, thereby enhancing emotional engagement as well as immersion in ideological and political learning. Ultimately, this approach fosters a deeper understanding among students regarding the content of ideological and political education.

3.3 Enhance Efficiency

With the advent of AI technology, the teaching methodology for ideological and political courses has undergone a fundamental transformation. Traditionally, these courses have been predominantly teacher-centered, resulting in a somewhat monotonous instructional approach that often fails to fully engage students' initiative in learning. However, through the analysis of learning behavior data, intelligent systems can accurately assess students' mastery of knowledge, learning preferences, and cognitive characteristics. This enables them to offer personalized learning suggestions and guidance tailored to individual needs, thereby facilitating differentiated teaching models [14]. Moreover, AI-driven systems can identify students' learning weaknesses based on assessment outcomes and intelligently recommend targeted exercises as well as supplementary learning materials. They are also capable of planning personalized learning pathways for each student while stimulating their interest in learning and intrinsic motivation. Additionally, features such as real-time Q&A and collaborative learning within intelligent interactive platforms foster deeper interactions between teachers and students. This "student-centered" teaching model not only effectively engages students' initiative in their educational pursuits but also enhances the overall effectiveness of education. It provides a robust foundation for cultivating individuals with strong ideals and beliefs alongside solid theoretical knowledge and practical skills.

4 Practical Challenges

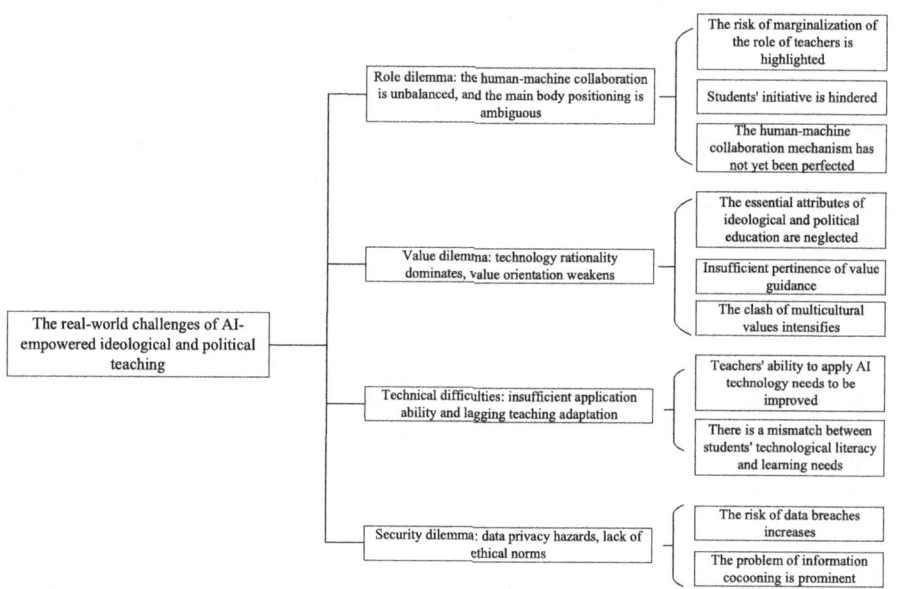

Fig. 2. Practical challenges of AI-empowered ideological and political teaching

4.1 Difficulties in Roles

In the process of deeply integrating artificial intelligence (AI) into the teaching of ideological and political courses, intelligent technology significantly impacts the traditional "teacher-led, student-centered" instructional model through functions such as algorithmic recommendations, data analysis, and virtual interactions. When educators utilize AI to assist in their teaching practices, there is a tendency to rely heavily on automated lesson plan generation and question-and-answer systems. This reliance can lead to an excessive standardization of teaching design and a gradual erosion of innovative pedagogical capabilities. This imbalance in human-machine collaboration results in teachers lacking essential educational functions that are irreplaceable, such as value guidance and emotional engagement. Furthermore, within the "information cocoon" created by

AI technologies, students often passively follow predetermined learning pathways. Consequently, their ability to think independently and actively explore new ideas is hindered [15]. This imbalance not only affects the dominant role of teachers but also hinders the development of students' initiative, making it challenging to achieve optimal effectiveness in human-machine collaboration. First, the risk of marginalizing the role of teachers is underscored. As artificial intelligence becomes increasingly integrated into the teaching of ideological and political courses, intelligent teaching systems can swiftly generate lesson plans and respond to student inquiries due to their advanced data processing capabilities and extensive knowledge reserves. This shift has significantly impacted the traditional function of teachers as knowledge transmitters. Some educators have become overly reliant on AI tools, transforming into mere "operators" of technology who directly utilize intelligently generated lesson plans and course materials. This reliance often leads to a neglect of secondary processing and value enhancement in teaching content, resulting in a lack of personal characteristics and emotional engagement from teachers during classroom instruction. Second, the initiative of students is hindered. The intelligent learning environment created by AI technology appears to offer students the convenience of independent learning; however, it may inadvertently inhibit their agency. On one hand, AI algorithms effectively push content based on students' historical learning data, resulting in a closed knowledge acquisition pathway where students passively accept predetermined learning resources. This leads to difficulties in accessing diverse perspectives and diminishes their motivation to actively explore and engage in critical thinking. On the other hand, the real-time feedback mechanism inherent in intelligent evaluation systems causes students to focus excessively on academic results and scores while neglecting self-reflection and skill development throughout the learning process. In ideological and political education, this lack of student initiative impedes their comprehensive understanding of theories as well as the internalization of values, making it challenging to fulfill the educational objectives aimed at shaping character and nurturing individuals. Third, the mechanism of human-machine collaboration has yet to be fully refined. Currently, the integration of artificial intelligence (AI) into ideological and political education remains in an exploratory phase, revealing numerous deficiencies within the human-machine collaboration framework. Teachers often struggle to determine the appropriate extent of AI assistance in their teaching practices, leading to instances of either excessive dependence on or outright rejection of technology. Furthermore, AI systems frequently fall short in addressing the specific needs inherent to ideological and political instruction; they exhibit notable limitations in emotional engagement and value-oriented guidance. As a result, the potential for maximizing efficiency through human-computer collaboration is not being realized effectively, which significantly hampers improvements in the quality of ideological and political course instruction (Fig. 2).

4.2 Value Dilemma

With the extensive application of AI technology in the teaching of ideological and political courses, the rational thinking inherent in algorithmic logic and data-driven methodologies has gradually permeated every aspect of instruction. This development presents a profound contradiction to the value orientation of "cultivating individuals with virtue" that underpins ideological and political education. In practical teaching scenarios, certain intelligent teaching platforms decompose ideological and political courses into knowledge point question banks and standardized assessments, excessively emphasizing quantitative metrics such as learning completion rates and accuracy rates. Consequently, this focus often neglects the essential cultivation of students' ideals, beliefs, and moral sentiments, leading to a significant weakening of the value-oriented function that is central to ideological and political courses [16]. The specific manifestations are as follows: First, the fundamental attributes of ideological and political education are overlooked. In the context of AI-enhanced ideological and political teaching, technical rationality assumes a predominant role, placing excessive emphasis on quantitative indicators of teaching efficiency and knowledge dissemination. This reductionist approach simplifies ideological and political courses to mere mechanical indoctrination of knowledge points, thereby neglecting the essential characteristics inherent in ideological and political education. Such a deviation from its core essence results in these courses losing their primary functions of providing ideological guidance and value orientation, making it challenging to cultivate students' correct worldview, outlook on life, and values. Second, there is an insufficiency in the relevance of value guidance. While AI leverages big data technology to facilitate personalized learning experiences, its algorithmic models predominantly rely on behavioral data. This reliance makes it difficult for these systems to penetrate beyond surface-level interactions to capture students' deeper cognitive fluctuations and emotional needs. Consequently, this technical limitation gives rise to what can be termed the "precision paradox" in value guidance: although intelligent systems may effectively deliver knowledge content, they fail to achieve an organic integration between emotional resonance and value directionality. As a result, they struggle to accommodate the diverse value requirements across different student demographics; thus leaving value guidance at a superficial and stylized level [17]. Finally, the clash of multicultural values intensifies. The digital transformation of education has exacerbated the complexity of value conflicts. Intelligent platforms offer extensive access to global information resources, which not only broadens avenues for knowledge acquisition but also facilitates the subtle infiltration of Western values such as individualism and consumerism through algorithmic recommendations. For instance, in an effort to enhance user engagement, certain AI recommendation algorithms present Western ideals like individualism and consumerism under the guise of "pop culture" to students. This practice undermines the development of sound values among students, complicates their ability to discern right from wrong amidst conflicting multicultural perspectives, and poses significant challenges for value guidance within ideological and political education courses.

4.3 Technical Difficulties

On the one hand, there is a pressing need to enhance teachers' proficiency in applying AI technology. Despite the growing popularity of AI applications within the educational sector, many ideological and political educators have not undergone systematic training in this area. Consequently, their familiarity with intelligent teaching tools remains limited, and their overall capability to utilize AI technology is still at a rudimentary level. This inadequacy prevents them from fully harnessing the advantages that AI can offer in teaching. Furthermore, teachers often possess a superficial understanding of how to integrate AI technology into ideological and political education. As a result, they face challenges when attempting to incorporate AI effectively into both instructional design and practical teaching scenarios. This shortcoming leads to instances where AI-enhanced ideological and political instruction becomes merely procedural or superficial, failing to achieve its intended outcomes. On the other hand, there exists a significant disparity between students' technological literacy and their learning needs. As a crucial component of AI-enhanced ideological and political education, students' technical proficiency is inconsistent, revealing a notable gap between their capabilities and the requirements for intelligent learning. Some students struggle with the operation of AI learning tools, hindering their ability to engage in self-directed learning effectively through intelligent platforms. Moreover, many students do not recognize the functionality and value of AI technology within the context of ideological and political education, which prevents them from fully realizing their academic potential. Additionally, there are variations in the learning needs across different student demographics; however, current AI teaching tools often lack personalized design features that could address these diverse requirements. This results in a misalignment between "technology supply" and "learning needs", ultimately impacting both learning effectiveness and teaching quality [18].

4.4 Security Difficulties

In the context of AI-enhanced ideological and political education, concerns regarding data security and ethical standards have become increasingly prominent. The risk of data privacy breaches is escalating, alongside the emergence of ethical dilemmas such as information cocooning. These issues collectively pose a significant threat to the healthy development of ideological and political teaching [19]. First, the risk of data breaches has escalated. In the context of AI-enabled ideological and political education, systems are required to collect and analyze substantial amounts of students' personal information, including sensitive data such as learning behaviors, ideological trends, and personal details. However, the current data security protection framework is inadequate; it not only faces risks from hacker attacks due to technical vulnerabilities but also harbors potential dangers of data leakage stemming from non-standard management practices. Second, the issue of information cocoons has become increasingly prominent. The algorithmic recommendation mechanism has given rise to what can be termed "information cocoon 2.0". The intelligent platform facilitates precise delivery of ideological and political education resources through user profiling. While this enhances the efficiency of information acquisition, it simultaneously creates a closed space for value recognition where students are exposed predominantly to singular and homogeneous information over extended periods. This technological monopoly results in students being immersed in a one-dimensional informational environment that restricts exposure to diverse ideas. Consequently, it impedes the development of critical thinking skills, exacerbates cognitive biases among students, complicates their ability to embrace differing perspectives and values, undermines their capacity for forming comprehensive and objective value judgments—ultimately conflicting with the overarching goal of cultivating well-rounded socialist builders within ideological and political curricula.

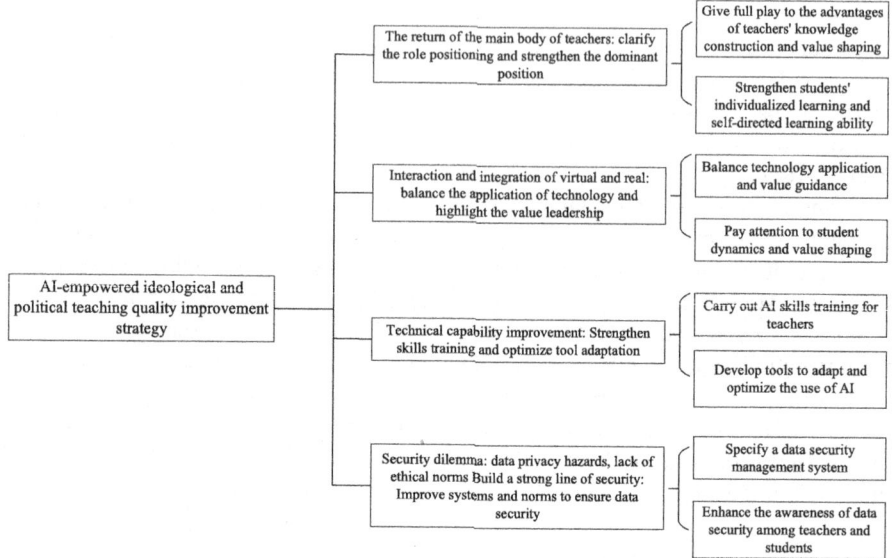

Fig. 3. AI-empowered ideological and political teaching quality improvement strategy

5 Optimize the Path

5.1 The Return of Teachers and Students

In the curriculum system of ideological and political education, teachers have consistently assumed dual roles as core organizers and value leaders, rendering their subjective status irreplaceable. While artificial intelligence technology has demonstrated robust capabilities in data processing and resource integration, it should fundamentally be regarded as an auxiliary teaching system rather than a primary teaching entity. Teachers must leverage their professional advantages in knowledge construction and value shaping while utilizing AI technology as a tool to enhance teaching design, rather than relinquishing complete control over instructional leadership. Taking the course "Basic Principles of Marxism" as an example, although intelligent systems can rapidly generate numerous academic documents and teaching cases, these materials may present issues such as theoretical interpretations that exceed students' cognitive development stages or case studies that are disconnected from the specific historical context and practical life experiences of the era. Therefore, teachers need to apply their professional judgment to redevelop AI-generated content by reconstructing theoretical frameworks using everyday language. They should also establish cognitive connections through current events and employ interactive Q&A sessions to stimulate critical thinking links [20]. For instance, when elucidating the "surplus value theory", educators can introduce a novel form of labor within the framework of the digital economy and guide students in comprehending the logic behind capital operations through AI algorithm recommendation mechanisms. In terms of value guidance, ideological and political courses serve not only as platforms for knowledge dissemination but, more crucially, as avenues for cultivating students' socialist core values while fostering an accurate worldview, outlook on life, and set of values. In today's era characterized by an explosion of information, AI-generated complex data and diverse perspectives proliferate rapidly; this includes erroneous and one-sided viewpoints. Educators should encourage students to engage in critical thinking regarding this information, teaching them how to discern right from wrong and good from evil. Furthermore, instructors ought to guide students in analyzing issues from various angles so that they may grasp the complexity and multifaceted nature of problems. Simultaneously, it is essential to help students establish sound values that enable them to maintain clarity when confronted with diverse information sources and prevent them from being misled by misguided ideologies (Fig. 3).

As the central focus of education, students must reclaim their primary role in the teaching of AI-empowered ideological and political courses. AI technology offers students opportunities for personalized learning, effectively dismantling the "one-size-fits-all" approach prevalent in traditional educational models. Students are empowered to select what they wish to learn and how they prefer to engage with that material, based on their individual learning pace, interests, hobbies, and styles. When utilizing AI learning platforms, it is essential for students to take initiative by setting their own learning objectives, choosing relevant content, and assessing their outcomes rather than merely adhering to the platform's predetermined pathways. Furthermore, the active engagement of students extends beyond independent selection of learning materials; it fundamentally encompasses the development of autonomous learning capabilities and critical thinking skills. Educators play a crucial role in guiding students on how to effectively leverage AI tools for their studies. They should instruct learners on how to discern and evaluate information provided by these technologies. Teachers ought to facilitate student analysis, evaluation, and integration of AI-generated information while encouraging them to consider issues from multiple perspectives. This comprehensive approach aims not only at enhancing students' information literacy but also at fostering robust critical thinking abilities.

5.2 Interaction and Integration of Virtual and Real

The teaching of ideological and political theory is undergoing innovative transformation driven by technology, with intelligent technologies such as VR/AR paving a new avenue for course instruction through the creation of three-dimensional virtual environments. Taking revolutionary history education as an example, within the teaching unit on the "Spirit of the Long March", educators can equip students with VR devices to facilitate an immersive learning experience that simulates extreme conditions faced by Red Army soldiers—such as snow-capped mountains and vast grasslands—as well as intense battle scenes where they broke through enemy blockades. This approach aims to shift traditional one-way knowledge transmission into a multi-sensory interactive experience, enabling students to transition from passive recipients to active participants. Consequently, this fosters an enhancement in cognitive processes from abstract comprehension to embodied experiences, thereby deepening students' memory retention and emotional connection to historical events. The pedagogical value of virtual reality technology is prominently demonstrated through its remarkable spatio-temporal extensibility. This capability transcends the physical constraints of traditional classrooms, allowing learners to access digitally restored historical scenes at any time. Furthermore, with the assistance of intelligent interactive systems, students can engage in dialogues and exchanges with virtual historical figures. This not only enriches the forms of classroom instruction but also fosters students' historical thinking and critical cognitive skills [21]. However, while fully leveraging the advantages of virtual technology in teaching, it is essential to uphold the core functions of ideological and political courses. As leaders in the educational process, teachers must engage in meticulous planning during the instructional design phase to ensure that the value orientation conveyed through both content and activities aligns closely with the objectives of these courses. In teaching historical events, beyond merely presenting their surface details, students should be guided to delve deeply into the underlying laws of historical development and the value logic inherent within them. This approach aims to foster an understanding of the ideals, beliefs, and qualities of will encapsulated in these events, ultimately cultivating students' emotional connections to their families and country as well as enhancing their national self-confidence and sense of social responsibility.

In today's intricate and pluralistic social information landscape, it is imperative for educators to maintain a keen awareness of social dynamics and the ideological currents among students when analyzing contemporary social issues. They should fully leverage their guiding roles to assist students in developing a scientific and rational value judgment system. For instance, during the integration of pressing social issues into classroom instruction, teachers can employ a variety of pedagogical approaches: on the one hand, they may utilize virtual technology to recreate event scenarios, allowing students to grasp the entirety of the situation more intuitively; on the other hand, they can actively organize in-depth discussion activities that engage students meaningfully [22]. In the course of the seminar, students will be guided to conduct a thorough analysis of the interests involved in the incident and the value orientations upheld by all parties. This approach aims to cultivate students' critical thinking skills and enhance their ability to make sound value judgments. Furthermore, educators should emphasize establishing a strong connection between virtual experiences and real life. They should guide students to gain a profound understanding of the practical application value of ideological and political theoretical knowledge, helping them realize that theoretical concepts are not isolated but are intricately linked to everyday life. Building on this foundation, teachers should encourage students to translate the values and concepts learned in class into tangible actions. Students are urged to start from the nuances of daily life and actively engage in various social practice activities, thereby deepening their understanding and practical application of these values through experience.

5.3 Improvement of Technical Capabilities

To address the challenges of inadequate application capabilities of AI technology and the slow adaptation of tools, it is essential to establish an integrated ecosystem encompassing "teaching, learning, research, and application". This should be pursued through both teacher training initiatives and product research and development efforts.

In the realm of teacher skills training, it is essential to establish a three-tiered training system encompassing "fundamental literacy development, enhancement of application abilities, and breakthroughs in innovative practices". This framework aims to address the developmental needs of teachers across various levels [23]. In the foundational training phase, a dual-track course titled "Technical Cognition and Basic Operation" has been developed specifically for young educators and groups with limited skills. This program employs the "Cognitive Workshop Micro Certification Assessment" mechanism to systematically impart knowledge on the operational specifications and technical principles of AI educational tools. It also includes practical training modules such as intelligent lesson plan generation and learning data visualization, thereby facilitating a comprehensive training process that spans from setting teaching goals to aligning teaching resources. Emphasizing the deep integration of AI with curriculum ideology and political education, we will implement a blended training program centered around "problem-oriented case analysis". In the application ability training stage, we emphasize the integration of artificial intelligence (AI) with ideological and political education. This involves implementing case-based teaching and workshops that are oriented around problem-solving. By deconstructing exemplary teaching cases, educators learn to utilize AI for analyzing learning conditions, identifying variations in students' comprehension of different theoretical modules through AI-generated cognitive portraits, and subsequently designing differentiated instructional plans to optimize teaching design. In the innovative training stage, teachers are encouraged to explore the application of cutting-edge technologies within ideological and political education. They engage in AI-driven virtual simulation practical teaching projects through collaborations between schools and enterprises as well as scientific research initiatives. This approach aims to enhance the effectiveness of ideological and political courses.

In the realm of tool adaptation and optimization, a product development mechanism characterized by "demand-oriented collaboration in R&D and dynamic optimization" has been established. Through the collection of questionnaires, classroom observations, and in-depth interviews, a "three-dimensional research matrix" system was developed to identify teaching challenges. This led to the creation of an intelligent lesson preparation system featuring knowledge association visualization capabilities tailored to the abstract characteristics of ideological and political course theory. The system automatically generates knowledge context maps for ideological and political education, thereby assisting educators in constructing a three-dimensional teaching framework encompassing "concepts, principles, and methodologies". Simultaneously, leveraging affective computing technology, we designed an interactive classroom tool capable of real-time identification of students' emotional tendencies. When the system detects that students are distracted or confused regarding specific viewpoints, it automatically suggests relevant case studies or initiates discussions to facilitate emotional engagement in teaching.

5.4 Build a Strong Line of Security Defense

As the primary entity responsible for data security management, schools and educational departments must establish a rigorous and comprehensive data security management system that encompasses all stages of the data life cycle, from the initial collection to final disposal and destruction. For instance, regarding data collection, it is essential to clearly define the scope of data gathering to ensure that only necessary information directly related to ideological and political course instruction—such as students' classroom performance metrics, homework completion rates, test scores, etc.—is utilized for analyzing learning progress and outcomes. Conversely, sensitive personal information about students, including detailed home addresses and family economic conditions, should be collected with utmost caution. Simultaneously, implementing a stringent data access management system is crucial for safeguarding data security. The access rights of various personnel must be accurately delineated and strictly controlled based on their roles and responsibilities within the institution. For example, only instructors teaching ideological and political courses should have access to students' academic performance records; school logistics managers should not possess such privileges. Furthermore, it is imperative to develop a robust data security contingency plan that anticipates potential security incidents comprehensively while outlining appropriate response strategies.

Enhancing the data security awareness of both teachers and students constitutes a crucial line of defense in safeguarding data integrity. Schools should implement diverse and ongoing training programs, as well as promotional activities focused on data security, to elucidate its critical importance and common risks associated with it. This approach aims to instill a deep understanding of data security principles among educators and learners alike. Simultaneously, establishing a dedicated lecture hall for data security protection training is essential. This venue can facilitate instruction on fundamental skills related to data protection, such as creating complex and secure passwords, regularly updating passwords, refraining from using AI teaching platforms over unsecured networks, and avoiding indiscriminate clicking on unknown links. Such initiatives are designed to elevate the overall awareness of data security practices among teachers and students. Furthermore, in the context of ideological and political education courses, integrating content related to data security can be executed seamlessly [24]. Teachers of ideological and political courses can employ case analysis, group discussions, and other methods to guide students in effectively managing personal information. They should make students aware of the serious consequences that may arise from the leakage of personal data. Educators must instruct students not to indiscriminately provide personal information on untrustworthy websites and platforms, nor should they easily disclose their account numbers and passwords to others. Furthermore, it is essential to enhance students' legal awareness by guiding them in understanding the laws and regulations surrounding data security. Teachers should assist students in establishing appropriate concepts regarding network security so that they can fully benefit from the conveniences offered by AI-enhanced ideological and political education. Ultimately, this will empower them to consciously protect both their own data as well as that of others.

6 Summary

In the context of the digital transformation of education, the profound integration of artificial intelligence with ideological and political theory courses has emerged as an essential strategy in the modernization of education. The empowerment provided by AI injects innovative vitality into traditional ideological and political curricula, demonstrating significant advantages in reprogramming teaching content, expanding educational spaces, and optimizing teaching efficiency. This development prompts educators to critically examine the value boundaries associated with technological empowerment from a more dialectical perspective. Looking ahead, educators specializing in ideological and political theory must strive to achieve a balance between technological advancements and the fundamental essence of education. They should consistently uphold their core responsibilities within these courses by embedding mainstream values into algorithmic recommendations. Furthermore, they can leverage virtual simulations to reinforce foundational aspects of red culture while fostering critical thinking skills within intelligent interactive environments. Ultimately, it is crucial to realize an organic unity between technology empowerment and value leadership, thereby providing robust support for cultivating a new generation capable of undertaking the monumental task of national rejuvenation.

Acknowledgements. This research was supported by Shenzhen Education Science Planning (Key Project); "Research on the Practical Path to Unify the Political and Academic Nature of Ideological and Political Courses in Higher Vocational Education through Discourse Pattern Transformation" (No.zdzz24027).

References

1. Wang, Y.K., Shen, X.R.: DeepSeek-like AI empowering ideological and political education in universities: value, concerns, and solutions. J. Southwest Petrol. Univ. (Soc. Sci. Edn.) **05**, 1–9 (2025)
2. Wang, Y.X., Liu, M.W.: AI-empowered precision teaching in college ideological and political education: technological foundations, risks, and coping strategies. Health Vocat. Educ. **43**(09), 50–54 (2025)
3. Yu, X.C., Yang, L.: Generative AI empowering college ideological and political education: prospects, risks, and pathways. China Educ. Technol. (04), 1–15 (2025)
4. Liu, S.Y.: AI-Empowered high-quality development of college ideological and political education: practical pathways. High. Educ. Forum (04), 10–18 (2025)
5. Xiao, F.Y.: AI-driven teaching reform in college ideological and political education: internal mechanisms, risk challenges, and countermeasures. E-Educ. Res. **46**(05), 103–107+115 (2025)
6. Liu, Y.J., Chen, Y.L.: Generative AI empowering ideological and political education: value implications, practical dilemmas, and implementation paths. J. Theory (04), 122–128 (2025)
7. Xing, M.: AI-empowered ideological and political education in colleges: current landscape, strategic directions, and innovative pathways. J. Chengdu Aeron. Polytechn. **41**(01), 103–108 (2025)
8. Zhou, C.Y.: Digital intelligence empowerment for vocational college teachers in ideological and political education: internal mechanisms and innovative approaches. J. Hubei Open Vocat. College **38**(08), 6–8 (2025)
9. Zhao, Y.Q., Wang, J.S., Wang, J.: A three-dimensional perspective on AI-empowered ideological and political education in higher education. J. Jinzhou Med. Univ. (Soc. Sci. Edn.) **23**(02), 97–99 (2025)
10. Liu, Y.J., Chen, Y.L.: Generative AI in ideological and political education: value implications, practical challenges, and implementation pathways. J. Theory (04), 122–128 (2025)
11. Xue, S.W: Exploring AI-empowered teaching approaches for college ideological and political theory courses. PR World (06), 196–198 (2025)
12. Lai, S.J., Feng, L.X.: The three-dimensional orientation of AI-empowered ideological and political education in universities. J. Soc. Sci. Jiamusi Univ. **43**(04), 166–169 (2025)

13. Ma, W.Q., Yang, X.L.: The triple dimensions of AI-empowered ideological and political education. China Educ. Technol. (04), 125–133 (2025)
14. Liu, H.B.: Digital technology-empowered ideological and political education in universities: practical logic, value implications, and implementation pathways. Henan Educ. (High. Educ.) **03**, 42–44 (2025)
15. Yi, G.Y.: Research on the reform of ideological and political education in universities from the perspective of AI. J. Changchun Normal Univ. **44**(03), 169–171 (2025)
16. Li, X.Z.: Exploring the pathways of digitally empowered ideological and political education in vocational colleges. J. Ezhou Univ. **32**(02), 26–28+72 (2025)
17. Xiong, F.B.: Smart classroom in college ideological and political education: technology-driven, integrated exploration, and innovative dimensions. Ideol. Theoretic. Educ. (03), 64–70 (2025)
18. Hu, Y., Wei, L.: Digital intelligence empowerment for high-quality development of ideological and political education in vocational colleges: an analysis of practical approaches. Zhongguancun (02), 186–188 (2025)
19. Zhang, M.H.: Digital intelligence technology empowering practical teaching in vocational college ideological and political education: value, challenges, and pathways. J. Guangdong Polytechn. Light Indust. **24**(01), 42–48 (2025)
20. Li, Y.C.: The urgency, practical challenges, and development pathways of integrating AI technology in college ideological and political education. J. Jinzhong Univ. **42**(01), 81–85 (2025)
21. Liao, L.: AIGC empowering teaching innovation for college ideological and political educators: opportunities, challenges, and countermeasures. Soc. Philanthropy (04), 283–286 (2025)
22. Zhu, W.X., Tao, L.: AI-empowered ideological and political education: teaching models, issues, and governance pathways. Party Build. Ideological Educ. Schools (04), 57–60 (2025)
23. Fang, Y.: AI-Empowered ideological and political education: value implications, practical pathways, and strategic exploration. Jiangsu Educ. Res. (02), 67–71 (2025)
24. Li, J.L., Zhou, Y.Y., Tang, R.: Directions·strategies·key aspects: research on the development trends of digitally empowered ideological and political education. J. Zhejiang Wanli Univ. **38**(01), 109–116 (2025)

Research on the Influence Mechanism of User-Generated Content Characteristics on Purchase Intention Under Big Data Recommendation

Ning Li, Chu Sun[✉], Yan Li, and Yongqi Ou

Guangdong University of Education, Guangzhou 510303, China
cissymix@163.com

Abstract. In the era of big data, the rapid development of intelligent recommendation technology has driven social e-commerce to become the most promising emerging business model. Among them, user-generated content (UGC), as an important carrier of information dissemination, has a great influence on stimulating potential users 'purchase intention. Therefore, this paper will focus on the three characteristics of user-generated content (UGC): interactivity, professionalism and authenticity, and deeply analyze its influence mechanism on consumers' purchase intention. Based on the SOR theory, the structural equation analysis method is used for empirical analysis, and the following conclusions are drawn: (1) UGC features have a significant positive impact on perceived value; (2) Perceived value has a significant positive impact on purchase intention; (3) UGC characteristics have a significant positive impact on consumer purchase intention. (4) In the process of UGC characteristics affecting purchase intention, perceived value plays an intermediary role. The above conclusions are of great help to e-commerce merchants to attach importance to the management of UGC and expand their own revenue.

Keywords: User-generated Content · Purchase Intention · Big Data Application

1 Introduction

With the rapid development of big data technology and mobile Internet technology, China 's e-commerce industry has entered a new stage of national e-commerce and intelligent e-commerce. According to the 54th data of the latest" China Internet Development Statistics Report "released by the China Internet Information Center, as of June 2024, the number of Internet users in China has reached nearly 1.1 billion, and the Internet penetration rate is as high as 78%. The number of network video users is 1.068 billion, accounting for 97% of the total number of Internet users. With the popularity of big data application technologies such as social media, e-commerce platforms, and intelligent recommendation and matching, consumers have changed from passive information receivers to active content producers, paying more attention to participation and sharing in the process of e-commerce consumption, that is, User-Generated Content

(UGC), which can form a fissionable communication through the social relationship network between users and significantly increase purchase intention. Enabled by big data technology, the value of UGC has gone beyond a single information transmission function. It has not only become the key basis for consumers' information acquisition and decision-making, but also become the core element driving enterprise decision-making and innovative marketing.

For enterprises, effectively mining the multi-modal features of UGC, such as text, image and video, can not only accurately identify consumer preferences, but also predict market trends through sentiment analysis, topic clustering and other technologies, so as to realize product innovation, personalized recommendation and brand word-of-mouth management. Closed-loop optimization, on the one hand, reduces marketing costs and improves conversion efficiency, on the other hand, promotes the value co-creation of brands and consumers by forming a positive cycle of 'content-data-insight-action' through user participation, highlighting the big data support provided by UGC for enterprises to insight into user needs and optimize marketing strategies. For example, Xiaohongshu, a well-known social e-commerce platform in China, combines UGC with big data intelligent recommendation system to enable its recommendation system, so that the products recommended to users are closer to consumers' interests and needs, thus improving the purchase conversion rate and successfully creating an intelligent community e-commerce ecology in the era of big data.

However, at present, the threshold and production cost of UGC are relatively low, the quality level is mixed, and the ability of big data recommendation is uneven. As a result, consumers are afraid to believe in UGC, which affects the transformation of user creativity into a new driving force for enterprise growth. Therefore, based on the background of big data era and from the perspective of consumers, this study analyzes what characteristics of UGC can attract consumers 'attention, win trust and generate corresponding purchase intention, so as to guide merchants and social platform providers to carry out more effective UGC management, better promote the performance growth and brand building of all parties, and provide theoretical support and practical path for enterprises to build a virtuous cycle of 'data-driven-UGC empowerment-business transformation'.

2 Literature Review

The full name of UGC is User Generated Content. Its concept was born in the Internet field and emerged with Web2.0. Liu Xuechun (2020) defines user-generated content as content that users post about product details, reviews, experiences, and purchase suggestions in the form of text, pictures, or videos on a platform that integrates trading and social networking. Zhao Yuxiang (2012) believes that user-generated content is a dynamic user creation behavior model, which is an order closely related to user groups, social networks, communication channels and online communities.

At present, the main directions of UGC research are classification, feature, motivation and effect research. Zhang Yihan (2014) divides the form of carrying content into text, pictures, video or audio created by users. Krishnamurthy (2008) divides users into two categories: rational users who share knowledge and thank users for social entertainment from the perspective of the purpose of participating in UGC activities. According to the number of participants, UGC is divided into two categories: group-generated content and individual-generated content. Wei Ruqing (2016) summarized UGC features as informative and normative. Xue Yunjian (2022) believes that UGC can be divided into richness, immediacy and reliability. Zhu Qinghua (2009) divides the main drivers affecting UGC into social-driven dimension, individual-driven dimension, technology-driven dimension and demographic characteristics adjustment set. Based on the two-factor theory, Sun Shaojun (2017) found that such incentive factors as reputation, entertainment benefits and self-efficacy will have a direct impact on the initiative of users to create content. Vasumathi (2024) found through a questionnaire survey that the information on social media has a significant impact on the choice of restaurants. It is expected that its influence will gradually increase with the advancement of big data technology. Li Meichan (2022) said that high-quality UGC can enhance users 'credibility of content. The clearer and more detailed UGC can enhance users' awareness of the authenticity of content, so as to trust content publishers and platforms more, and then promote the generation of purchase intention.

In summary, most of the existing research is based on the quality and impact of generated content, and less explores the typical social characteristics of user-generated content, and lacks the influence of UGC features under intelligent recommendation in the era of big data. Therefore, through the big data recommendation of UGC features, the division is deconstructed in the social e-commerce scene, and the influence mechanism of UGC features on promoting consumers 'purchase intention is studied, which expands the richness of UGC research theory system.

3 Theoretical Framework

3.1 UGC Features and Perceived Value

Yu (2021) considers UGC to be interactive, Muda M (2021) considers UGC to be authentic, Liu Xuechun (2020) believes that professional content can allow users to better understand the information of the product, the higher the recognition of the product, and stimulate the user's emotional response. Wang Jianing (2022) found that users can more intuitively measure and perceive the value and service of goods by searching for the details of goods and real shopping experiences shared by other users through social platforms. Ased on the above, the following assumptions are made:

H1: Interactivity significantly positively impact on perceived value;
H2: Professionalism has a significant positive impact on perceived value;
H3: Authenticity has a significant positive impact on perceived value.

3.2 Perceived Value and Purchase Intention

Zeithaml (1988) pointed out that perceived value is the overall evaluation formed by consumers after weighing the gains and losses of the utility of a product, which reflects the comparison between the actual satisfaction and expectations of consumers after paying the cost. Accordingly, this paper defines perceived value as the actual value that users feel about products or services in social e-commerce. Shahzeb (2022) empirically found that multiple independent factors based on social media in e-commerce may affect online purchase behavior, and perceived value as a mediator is not as significant as the role of social networking sites themselves. Li Meichan (2022) found that the higher the user's perceived value, the more it can improve their purchase intention. Based on the above, the following hypotheses are proposed:

H4: Perceived value has a significant positive impact on purchase intention;

H5: In the process of UGC characteristics affecting purchase intention, perceived value plays an intermediary role.

3.3 UGC Characteristics and Purchase Intention

Shen Mengjie (2022) defined purchase intention as the tendency to purchase goods or services based on consumers' subjective understanding. Tian Jiaxin (2023) defines purchase intention as the possibility of purchasing a product after the user has a certain degree of understanding of the goods or services by browsing text, pictures or videos. Based on this, this paper defines the purchase intention as the possibility of consumer' purchase tendency of specific goods based on personal or external subjective cognition. Wang Jianing (2022) believes that detailed content sharing and interactive environments, relatively real and effective feedback comments, valuable information, and the potential for sharing can all increase the likelihood of purchase. Ye Zi (2023) found that high-quality content with authenticity and content published by interactive content publishers will positively affect consumers' purchase intention. In summary, the following assumptions are made:

H6: Interaction has a significant positive impact on purchase intention;

H7: Professionalism has a significant positive impact on purchase intention;

H8: Authenticity has a significant positive impact on purchase intention.

3.4 S-O-R Theory

The S-O-R theory is the 'stimulus-organism-response' theory. The stimulus variable (S) is an external environmental factor that may change consumers 'cognition and emotions when purchasing goods. The body variable (O) is a process of individual psychological change between stimulation and final response. The response variable (R) is the final response of consumers to changes in their perceptual emotions and psychological activities after being stimulated by the external environment. It was first proposed by Mehrabian and Russell in 1974, and was first used to explain the influence of the external environment on people's psychological activities and behaviors. Later, it was gradually widely used to study consumer behavior. The S-O-R model indicates that consumers are stimulated by various external environments, and consumers generate purchase behavior through individual perception. Therefore, the S-O-R theory is very suitable as an important theoretical basis for studying whether the user-generated content in the social e-commerce platform, that is, the text, pictures, video and other information released by the user, can affect the consumer's purchase intention.

Based on the above assumptions, a theoretical model diagram is constructed based on the S-O-R theory, as shown in Fig. 1.

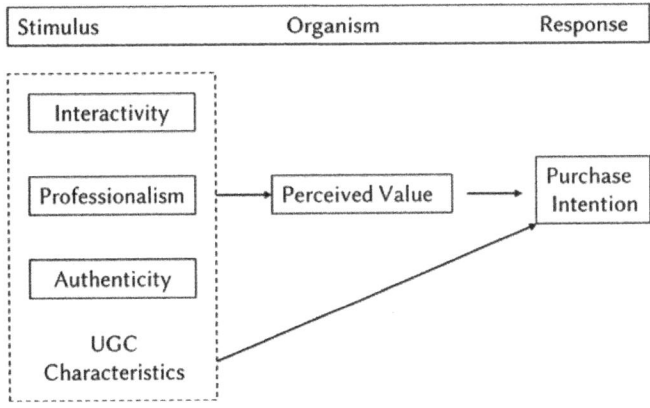

Fig. 1. Theoretical Model Framework

4 Empirical Analysis

The questionnaire uses the Likert five-level scale method to quantify the items, which is divided into five levels: very disagree, disagree, general, agree, very agree. The score is from 1–5, and the larger the number is, the more consistent the situation is. In terms of the source of the item, the UGC characteristic variables mainly refer to the research of Tan Shunxin (2022), Shen Mengjie (2022). The variable of perceived value refers to the results of Dai Xuchen (2022) and Wu Jiaojiao (2022). The purchase intention variable refers to the research of Geng (2021) and Liu Xuechun (2020).

The questionnaire is made by means of the questionnaire star platform, and the questionnaire link and two-dimensional code are distributed through the social e-commerce platform. After 15 days of distribution and collation, a total of 258 questionnaire samples were collected. After removing 55 invalid questionnaires, a total of 203 valid questionnaires remained, accounting for 78.68% of all questionnaires.

Table 1. Descriptive statistical analysis

Variable	Option	Numbers	Percentage
Gender	male	101	49.8%
	female	102	50.2%
Age	20 or less	17	8.4%
	21–30	88	43.3%
	31–40	74	36.5%
	41–50	13	6.4%
	Over 51	11	5.4%
Education	high school	37	18.2%
	college degree	146	71.9%

(*continued*)

Table 1. (*continued*)

Variable	Option	Numbers	Percentage
	master degree	20	9.9%
Income(mouth)	Under 2000	26	12.8%
	2000–4000	24	11.8%
	4001–6000	61	30.0%
	6001–8000	75	36.9%
	Above 8000	17	8.4%

The distribution of descriptive statistics of this questionnaire survey is shown in Table 1. The distribution of the survey samples is more balanced under various indicators.

Table 2. Reliability Analysis

	Cronbach 's α	Item No.
Interactivity	0.914	HD1-HD7
Professionalism	0.909	ZY1-ZY7
Authenticity	0.8	ZS1-ZS3
Perceived value	0.915	JZ1-JZ7
Purchase Intention	0.92	GM1-GM7

Table 3. Validity analysis

Variable	Item	Factor loads	Anova proportion	AVE	CR	KMO	Bartlett
Purchase Intention	GM5	0.806	15.35%	0.623	0.920	0.925	0.000
	GM2	0.785					
	GM6	0.777					
	GM1	0.772					
	GM7	0.764					
	GM4	0.743					
	GM3	0.727					
Perceived value	JZ1	0.808	30.68%	0.608	0.916		
	JZ7	0.792					
	JZ3	0.787					
	JZ6	0.784					
	JZ5	0.778					

(*continued*)

Table 3. (*continued*)

Variable	Item	Factor loads	Anova proportion	AVE	CR	KMO	Bartlett
	JZ2	0.77					
	JZ4	0.728					
Interactivity	HD4	0.787	45.78%	0.603	0.914		
	HD5	0.777					
	HD3	0.77					
	HD1	0.766					
	HD7	0.764					
	HD6	0.763					
	HD2	0.737					
Professionalism	ZY6	0.788	60.64%	0.588	0.909		
	ZY5	0.78					
	ZY3	0.772					
	ZY4	0.768					
	ZY2	0.768					
	ZY7	0.743					
	ZY1	0.73					
Authenticity	ZS2	0.798	67.36%	0.577	0.803		
	ZS3	0.774					
	ZS1	0.714					

After using SPSS software to analyze the reliability and validity, the results of the analysis are shown in Tables 2 and 3. It was found that the Cronbach's α coefficient values of UGC characteristics, perceived value and purchase intention were all greater than 0.8, and the reliability test was qualified. The overall KMO value was greater than 0.7, and the Bartlett spherical test P value was less than 0.05, which was suitable for exploratory factor analysis. In the principal component analysis, after using the maximum variance method to rotate, the cumulative variance interpretation rate of the extracted total factor was 67.36%, and the factor loads under each item were greater than 0.5, indicating that the structural validity was good. The AVE values of each variable dimension were more than 0.5, and the CR values were more than 0.7, indicating that the convergent validity was good.

Table 4. Confirmatory factor analysis

Statistical test	Output	Reference value	Result
χ^2/df	1.035	1–3 perfect; <5 good	perfect
RMSEA	0.013	<0.05 perfect; <0.08 good	perfect

(*continued*)

Table 4. (continued)

Statistical test	Output	Reference value	Result
AGFI	0.865	>0.80	accept
GFI	0.885	>0.80	accept
NFI	0.894	>0.90	accept
RFI	0.884	>0.90	accept
IFI	0.996	>0.90	perfect
TLI	0.996	>0.90	perfect
CFI	0.996	>0.90	perfect
PGFI	0.756	>0.50	perfect
PCFI	0.908	>0.50	perfect
PNFI	0.815	>0.50	perfect

With the help of AMOS software, the fitting degree analysis was carried out in Table 4, χ^2/df was 1.035, RMSEA was 0.013, NFI was 0.894, RFI was 0.884, TLI was 0.996, PGFI was 0.756, and each index was within the acceptable range, indicating that the fitting degree of the model and the actual survey data was up to standard.

Table 5. Multiple Regression Analysis

Model			Standardized Coefficient	t	Significance Level	Collinearity	
						tolerance	VIF
Dependent Variable: Perceived value	Constant		/	4.30	0	/	/
	Independent Variable	Interactivity	0.139	1.98	0.049	0.775	1.29
		Professionalism	0.223	3.19	0.002	0.78	1.282
		Authenticity	0.266	3.74	0	0.756	1.323
	Adjusted R-squared				0.231		
	F				21.218***		
	D-W				1.89		
Dependent Variable: Purchase Intention	Constant		/	3.08	0.002	/	/
	Independent Variable	Interactivity	0.331	5	0	0.775	1.29
		Professionalism	0.183	2.77	0.006	0.78	1.282
		Authenticity	0.215	3.207	0.002	0.756	1.323
	Adjusted R-squared				0.315		
	F				32.018***		
	D-W				1.978		
Dependent Variable: Purchase Intention	Constant		/	8.731	0	/	/
	Independent Variable	Perceived value	0.39	5.998	0	1	1

(continued)

Table 5. (*continued*)

Model		Standardized Coefficient	t	Significance Level	Collinearity	
					tolerance	VIF
	Adjusted R-squared			0.148		
	F			35.976***		
	D-W			1.997		

After three groups of multiple regression analysis, the data results presented in Table 5 were obtained. After analysis, it was found that the adjusted R and D-W values were within a reasonable range, the VIF was less than 5, and the F test was also within the significant level range, indicating that the explanatory power of the regression analysis model was relatively strong.

The dependent variable is the column of perceived value. It can be seen that the p values of the three independent variables are all less than 0.05, and the standardized coefficients are all greater than 0. It can be seen that the influence of interactivity, professionalism and authenticity on perceived value is significant and positive, so H1, H2 and H3 are verified. The dependent variable is the purchase intention, and the independent variable is the column of interaction, professionalism and authenticity. It can be seen that the p values of the three independent variables are all less than 0.05, and the standardized coefficients are all greater than 0, which proves that the interaction, professionalism and authenticity have a significant positive impact on the purchase intention. Therefore, H6, H7 and H8 are verified. The dependent variable is the purchase intention, and the independent variable is the column of perceived value. The p value is less than 0.05, and the standardization coefficient is greater than 0. The impact of perceived value on purchase intention is significantly positive, which verifies Hypothesis H4.

Using the Process plug-in Model4 model in SPSS software, 5000 samplings and 95% confidence levels were set. When the independent variables were interactive, professional and authentic, three mediating effects were tested. If the upper and lower limits of the results did not include 0, the mediating effect was significant. In Table 6, three different independent variables do not contain 0 value in the confidence interval of 95% confidence level, and the indirect effects all have a certain proportion, indicating that perceived value is a partial mediating effect in the three groups of tests, verifying H5. At this point, all the research hypotheses listed in this paper have been supported and verified.

Table 6. Mediating Effect test

Independent Variable		Effect	LLCI	ULCI	Percentage Effect
Interactivity	Total effect	0.5031	0.3783	0.628	/
	Direct effect	0.4156	0.288	0.5433	83%
	Indirect effect	0.0875	0.0384	0.147	17%
Professionalism	Total effect	0.4261	0.2887	0.5635	/
	Direct effect	0.3109	0.1677	0.4541	73%

(*continued*)

Table 6. (*continued*)

Independent Variable		Effect	LLCI	ULCI	Percentage Effect
	Indirect effect	0.1152	0.0514	0.1895	27%
Authenticity	Total effect	0.4378	0.3085	0.5672	/
	Direct effect	0.3283	0.1906	0.4659	75%
	Indirect effect	0.1096	0.0481	0.1806	25%

5 Conclusion and Suggestion

5.1 Conclusion

Based on the analysis of the above research hypothesis results, the following conclusions are drawn.

The interactivity, professionalism, and authenticity of User-Generated Content (UGC) all have a significant positive impact on purchase intention. The conclusion of authenticity and interactivity is consistent with previous studies, but the positive influence relationship between UGC professionalism and purchase intention is different from other people's perspectives. This indicates that when there is a good interactive atmosphere on the social e-commerce platform, the user content generator has rich purchasing experience and professional knowledge, the content shared is clear and detailed, the information is accurate and credible, and the angle is objective and fair, which can effectively promote the generation of purchase intention.

The positive impact of interactivity, professionalism and authenticity of UGC on perceived value has been effectively verified. In the process of interaction, users can feel that they are not alone, they can share experiences and views with other users, and get support and recognition from their feedback. At the same time, professional UGC is often able to provide more accurate, more in-depth, more comprehensive information, more trustworthy. In addition, when users find that the information provided by UGC is accurate, they will have higher recognition and trust in the content of UGC. These feelings ultimately make users have perceived value for UGC.

Perceived value has a significant positive impact on purchase intention, and plays a partial intermediary role in the process of UGC characteristics affecting purchase intention. When the interactivity, professionalism and effectiveness of UGC are perceived by consumers, the stronger the perceived value of the product is, the more it can be stimulated. When the perceived value of the product is higher, it will reduce the doubts in the heart, recognize the value points contained in the product, and promote the user to have a stronger purchase intention.

5.2 Suggestion

The platform should set up a measurement mechanism from different dimensions of interactivity, professionalism and authenticity, establish a content audit management system, and strengthen the control of UGC content quality. In addition to manual audit, it should also strengthen technical identification methods and improve efficiency. In addition, we should improve the supervision service system, give full play to the initiative of users to supervise and report, increase the complaint reward feedback system, and stimulate the enthusiasm of users. For example, byte beating group, in the audit of the daily mass of new UGC content, the use of the first machine screening, some illegal prohibited words appear in a timely manner to delete the content, ease the pressure of manual audit. If there are some obscure violations of the content is presented, once the user found and complained, artificial rapid intervention, to verify the correctness, immediately clear, and to provide clues to the user advertising products or discount vouchers reward.

The platform can give high-quality content creators appropriate incentives, such as increasing traffic, gift boxes, medals, etc. It is also possible to hold contests for the public and elect favorite high-quality content creators, which not only increases user activity, but also explores more high-quality creators. For the majority of small and medium-sized enterprises, we can set up the UGC evaluation matrix tool at the company level to formulate more detailed evaluation rules for professionalism, authenticity and interactivity. For example, professionalism mainly evaluates the correlation degree and viewing time of UGC and competitive products, authenticity mainly evaluates the proportion of boutique UGC in the account, and interactivity mainly evaluates the number of messages and likes under UGC content. Through the assessment of the corresponding performance, employees are encouraged to publish more works that are useful to consumers and stimulate users to generate consumer demand.

The platform should actively create a good interactive atmosphere, regularly organize creative activities, attract users to participate in interactive sharing, encourage users to freely express and exchange experience, but resolutely crack down on threatening, offensive and vulgar remarks. For example, the Douyin platform, in addition to spreading the hottest topics in the current society, also actively constructs some creative topics, and guides the users of the whole platform to participate in the UGC creation of related topics through the form of bonus sharing, which further promotes the Douyin platform to retain active users. Small and medium-sized enterprises should take the initiative to connect with major social platforms, actively combine their own product characteristics with platform creation orientation, effectively improve the contact frequency of target customers, and promote the widespread dissemination of their own product image.

Small and medium-sized enterprises should use big data analysis technology to conduct in-depth research on the browsing content and access of target customers, conduct targeted guidance around their needs and preferences, improve the authenticity of UGC content related to their products, and launch more interactive UGC content. For example, PESSI Group focuses on the audience of key competition events every year. On the Tik Tok platform, it guides users to publish product creative videos, and uses data to promote services, targeted delivery of relevant UGC content, and continues to maintain its head brand position in the beverage industry. In addition, small and medium-sized enterprises use big data technology to analyze and process users 'comments and discussions on their goods and services on social media in a timely manner, so as to detect public opinion crisis as early as possible, and then intervene, which helps to enhance consumers' perceived value and trust.

Acknowledgments. This study was supported by the "Research on the reform of evidence-based teaching quality evaluation of Artificial Intelligence empowerment" (Project No. 2022jxgg26), Guangdong Higher Education Teaching Reform Project "Research on the reform of AI enabling evidence-based teaching quality evaluation and the construction of evaluation resource pool" and The Fourteenth Five-Year Plan for the Development of Philosophy and Social Science in Guangzhou Project "Study on Accelerating the Construction of a Strong Education City in Guangzhou—Teaching Quality Evaluation and Resource Pool Construction in Universities"(Project No.2023GZQN49), youth philosophy and Social Sciences in Guangdong project "Artificial Intelligence Enabled Evidence-Based Evaluation of Teaching Effectiveness in Universities in the Greater Bay Area" (Project No.GD23YJY06).Research Platforms and Projects of Guangdong Provincial Department of Education—Young Innovative Talents Project "Research on the Mechanism of Value Co-creation of E-commerce Digital Brand Community Empowered by Gamification in the Context of High Quality Development"(Project No.2023WQNCX052).

References

Liu, X.: A study on the impact of user-generated content characteristics on consumers' purchase intentions in a social e-commerce context. MA Thesis. Anhui University of Finance and Economics, China (2020)

Zhao, Y., Fan, Z., Zhu, Q.: Conceptualization and research progress on user-generated content. J. Lib. Sci. China **38**(5), 68–81 (2012)

Zhang, Y.: Analyze the connotation of UGC, explore the application of UGC:a review of research and application of user generated content of new generation under the internet environment. Lib. Inform. Serv. **58**(20), 145–148 (2014)

Krishnamurthy, S., Wenyu, D.: Note from special issue editors: advertising with user-generated content: a framework and research Agenda. J. Interact. Advert. **8**(2), 1–4 (2008)

Wei, R., Tang, F.: The social influencing mechanism of user-generated content on online purchasing-an empirical study based on social e-commerce platform. East China Econ. Manage. **30**(04), 124–131 (2016)

Zhu, Q., Zhao, Y.: Exploration of the motivations to produce user-generated content in web 2.0 era. J. Lib. Sci. China **35**(5), 107–116 (2009)

Sun, S., Zhang, Y.: Research on user participation motivation of socialized e-commerce UGC platform—taking xiaohongshu as an example. Design **262**(7), 14–15 (2017)

Vasumathi, A., Ambrose, J.: A study on the factors that impact consumer decision making process in the context of using social media for choosing a hotel in India among students. Int. J. Bus. Process Integrat. Manage. **11**(3), 221–236 (2024)

Li, M.: Research on influence factors of purchase intention of social e-commercial platform users: a case study of Xiaohongshu. China Circul. Econ. **2339**(35), 12–15 (2022)

Yu, J., Ko, E.: UGC attributes and effects: implication for luxury brand advertising. Int. J. Advert. **40**(6), 945–967 (2021)

Muda, M., Hamzah, M.I.: Should I suggest this YouTube clip? The impact of UGC source credibility on eWOM and purchase intention. J. Res. Interact. Market. **15**(3), 441–459 (2021)

Wang, J.: The impact of social-media UGC on its user purchase behavior. China Collect. Econ. **697**(05), 59–60 (2022)

Zeithaml, V.A.: Consumer perceptions of price, quality, and value: a means-end model and synthesis of evidence. J. Market. **52**(3), 2–22 (1988)

Zulfiqar, S., Ahmed, F.: Direct and mediating effects of social media on e-commerce purchase intention: a comparative approach. Int. J. Bus. Process Integrat. Manage. **11**(1), 19–25 (2022)

Shen, M.: A study of the impact of user-generated content on consumer purchase intention in social e-commerce. MA Thesis. Central China Normal University, China (2022)

Tian, J.: Research on the influencing factors of consumers ' purchase intention under user-generated content mode. Indust. Innovat. **120**(19), 75–77 (2023)

Ye, Z.: Research on the influence of user generated brand content characteristics on consumer purchase intention in UGC community industry. MA Thesis. Xi'an University of Technology, China (2023)

Tan, S.: Research on the impact of UGC characteristics of social e-commerce community on consumers' purchase intention—analysis based on "Xiaohongshu". MA Thesis. Southwest University of Finance and Economics, China (2022)

Dai, X.: a study of the impact of social media user-generated content on consumers' unplanned purchase behavior. MA Thesis. Shanghai University of Finance and Economics, China (2022)

Wu, J.: Research on the influence of user-generated content on consumers' purchase intention in social platforms. MA Thesis. Southeast University, China (2021)

Geng, R., Chen, J.: The influencing mechanism of interaction quality of UGC on consumers' purchase intention–an empirical analysis. Front. Psychol. **12**, 697382 (2021)

Big Data Analysis of Inflammatory Bowel Disease-Associated Autoantibodies in China

Xufu Xiang[1,2(✉)], Weifang Li[1], Xiaotao Lin[1], Gang Wang[1], and Chungen Qian[1,2(✉)]

[1] Department of Reagent Research and Development, Shenzhen YHLO Biotech Co., Ltd.,
Shenzhen, Guangdong, China
{liweifang,linxiaotao,wanggang,gen.qian}@szyhlo.com

[2] The Key Laboratory for Biomedical Photonics of MOE at Wuhan National Laboratory for Optoelectronics-Hubei Bioinformatics and Molecular Imaging Key Laboratory, Systems Biology Theme, Department of Biomedical Engineering, College of Life Science and Technology, Huazhong University of Science and Technology, Wuhan 430074, China
xiangxufu@hust.edu.cn

Abstract. The rising incidence of ulcerative colitis (UC) in China highlights the need for effective biomarkers. Autoantibodies targeting integrin αvβ6 show association with UC. This study developed a novel chemiluminescent immunoassay (CLIA) for detecting anti-αvβ6 autoantibodies by conjugating recombinant αvβ6 protein to superparamagnetic microparticles and detecting captured IgG with acridinium ester-labeled anti-human IgG. The assay was rigorously validated using 961 serum samples from China. Cutoff values were established using 186 UC patients and 244 healthy controls, while clinical sensitivity was assessed on 531 samples. The assay achieved a sensitivity of 72.7% and specificity of 94.7% (AUC = 0.846) for UC diagnosis, demonstrating exceptional specificity: 100% in children (n = 128) and pregnant women (n ≈ 171), 99.36% in adults (n = 132), and 99.00% in the elderly (n = 100), yielding an overall specificity of 99.35%. These results confirm the high sensitivity and excellent specificity of this CLIA for UC detection, aligning with global findings. Although further research into antibody titer correlations with disease severity is warranted, this assay demonstrates significant potential for clinical application as an auxiliary diagnostic tool for UC.

Keywords: αvβ6 integrin · Ulcerative colitis · Chinese population · Autoimmunity · anti-αvβ6 IgG

1 Introduction

Ulcerative colitis (UC), a chronic inflammatory bowel disease (IBD) characterized by diffuse mucosal inflammation and ulceration primarily affecting the colon and rectum, presents a growing clinical challenge in China. Its incidence and prevalence exhibit a marked upward trajectory, closely associated with rapid socioeconomic development and

---

Xufu Xiang and Weifang Li—These authors contributed equally

shifts in lifestyle and environmental factors [1–3]. UC typically manifests with persistent diarrhea, hematochezia, abdominal pain, tenesmus, urgency, fatigue, and weight loss, significantly impairing patients' quality of life and productivity [1, 4]. Striking predominantly young adults during their peak productive years, UC follows a chronic, relapsing course with substantial long-term morbidity. Complications range from acute severe colitis and toxic megacolon to an elevated risk of colorectal dysplasia and carcinoma, highlighting its profound clinical burden. Consequently, early and precise diagnosis is paramount for initiating timely interventions, mitigating disease progression, preventing complications, and improving patient outcomes.

The current diagnostic framework for UC integrates clinical assessment, endoscopic evaluation with histopathology, and supportive serological and radiological findings. Colonoscopy remains the indispensable gold standard, enabling direct mucosal visualization, assessment of disease extent, and acquisition of biopsies crucial for differential diagnosis, particularly distinguishing UC from Crohn's disease. However, its invasive nature, associated patient discomfort, need for bowel preparation, procedural risks, and resource intensity limit its utility, especially for repeated assessments. Non-invasive serological biomarkers, including C-reactive protein (CRP)[5], erythrocyte sedimentation rate (ESR) [6], perinuclear anti-neutrophil cytoplasmic antibodies (pANCA) [7, 8], and anti-Saccharomyces cerevisiae antibodies (ASCA) [9], serve as adjuncts. Nevertheless, their diagnostic performance is suboptimal. CRP and ESR reflect systemic inflammation but lack UC specificity. pANCA offers moderate specificity versus Crohn's disease but exhibits variable sensitivity and cross-reactivity with other autoimmune conditions. ASCA is more associated with Crohn's disease. This persistent gap between diagnostic accuracy, patient acceptability, and practicality underscores the critical need for novel, highly specific, and sensitive biomarkers to facilitate earlier diagnosis and reduce reliance on invasive procedures.

Integrins, heterodimeric (α/β) cell surface receptors, mediate vital cell-extracellular matrix (ECM) interactions governing adhesion, signaling, and cellular functions[10]. The $\alpha v \beta 6$ integrin, possessing restricted epithelial expression primarily during development, wound healing, and pathology, has garnered significant interest in UC pathogenesis [11]. Dysregulated $\alpha v \beta 6$ expression on intestinal epithelium contributes to mucosal inflammation via aberrant activation of latent transforming growth factor-beta (TGF-β) and modulation of epithelial barrier integrity [12, 13]. Crucially, the discovery of circulating autoantibodies targeting $\alpha v \beta 6$ (anti-$\alpha v \beta 6$ IgG) represents a major advance in UC serology. Seminal Japanese studies demonstrated exceptional diagnostic potential: anti-$\alpha v \beta 6$ IgG positivity occurred in 92% of UC patients (n = 112) versus only 5% in non-IBD controls (n = 155). Subsequent validation confirmed high clinical specificity, with significantly lower positivity rates in Crohn's disease (32.6%), IBD-unclassified (20%), and primary immunodeficiency (8%) compared to UC (94.7%). These findings firmly establish anti-$\alpha v \beta 6$ IgG as a highly promising, UC-specific serological biomarker[15, 16]. However, critical gaps hinder its clinical translation. Validation data within large, ethnically distinct populations like China are scarce. Furthermore, the lack of standardized, robust, and clinically applicable detection methodologies suitable for routine diagnostics represents a significant barrier.

Chemiluminescence immunoassay (CLIA) technology, particularly utilizing superparamagnetic microparticles, constitutes a major advancement in automated, high-performance in vitro diagnostics. Magnetic particle-based CLIA offers distinct advantages over ELISA, including superior sensitivity and specificity, broader dynamic range, faster turnaround times, full automation enhancing reproducibility, and reduced manual error. These attributes make it ideal for clinical laboratories and large-scale screening. Addressing the identified gaps, we developed a novel chemiluminescent immunoassay leveraging recombinant αvβ6 protein conjugated to superparamagnetic microparticles for specific autoantibody capture, with sensitive detection using acridinium ester-labeled anti-human IgG. Recognizing potential population-specific variations in immune responses, we established the first reference ranges for anti-αvβ6 IgG using a well-defined cohort from China. This study aims to rigorously evaluate the diagnostic performance (sensitivity, specificity) of this novel CLIA for detecting anti-αvβ6 IgG across diverse demographic groups within this Chinese population. Our work seeks to validate this promising biomarker in a critical, underrepresented cohort and provide a robust methodological platform, thereby facilitating its potential integration into clinical practice for improved UC diagnosis.

2 Materials and Methods

2.1 Study Cohorts and Sample Processing

Serum samples were obtained from Shenzhen Bokang Biotechnology Co., Ltd. (Shenzhen, China). For establishing diagnostic thresholds, a cohort comprising 430 individuals was analyzed: 243 healthy controls (ages 18–60; 119 male, 125 female) and 187 ulcerative colitis (UC) patients (ages 21–60; 103 male, 83 female). Clinical specificity was evaluated using an independent cohort of 531 subjects, stratified into pediatric (<18 years, n = 128), adult (18–60 years, n = 132), elderly (>60 years, n = 100), and pregnant (20–43 years, n = 171) subpopulations. Additionally, 68 rheumatoid factor (RF) IgM-positive samples (detected via iFlash-RF IgM kit [Shenzhen YHLO Biotech], ages > 18) were acquired for interference studies. All samples underwent standardized processing: clotting completion followed by centrifugation at 3,000 × g for 15 min, with aliquots stored at −20 °C if not analyzed immediately.

2.2 Anti-αvβ6 IgG Chemiluminescent Immunoassay

Anti-αvβ6 IgG detection was performed using a paramagnetic particle-based chemiluminescent immunoassay (CLIA) reagent (Shenzhen YHLO Biotech) on the iFlash automated platform. Recombinant human αvβ6 protein (R&D Systems) was covalently conjugated to carboxylated paramagnetic microparticles (Thermo Fisher Scientific) via N-ethyl-N'-(3-dimethylaminopropyl) carbodiimide (EDAC) chemistry (Thermo Fisher Scientific). A mouse anti-human IgG monoclonal antibody (Fapon Biotech) was labeled with NSP-DMAE-NHS ester (Maxchemtech) and purified by Sephadex G-50 gel filtration. The assay protocol involved: 1) incubating 5μL sample with αvβ6-conjugated microparticles for 10 min, 2) washing, 3) reacting with acridinium ester-labeled detection antibody for 10 min, 4) washing, and 5) triggering chemiluminescence with hydrogen peroxide. Signal emission at 340 nm was quantified, with results calibrated against a master curve and reported in AU/mL.

2.3 Assay Calibration and Analytical Validation

The diagnostic threshold was established by receiver operating characteristic (ROC) analysis of 430 samples (244 healthy, 186 UC). The optimal cutoff (20 AU/mL) maximized the Youden index, yielding an AUC of 0.846 with 95% CI. A calibrated master curve was generated using serial dilutions referenced to this cutoff. Precision was rigorously assessed per CLSI EP05-A3: triplicate samples (negative, positive, high-positive) were analyzed in duplicate twice daily for 20 days. Repeatability (within-run CV) and within-laboratory precision (total CV) were calculated from variance components. Linearity was evaluated across the analytical measurement range (3–4000 AU/mL) using six dilutions of high/low concentration pools (5H, 4H+1L, 3H+2L, 2H+3L, 1H+4L, 5L) analyzed in duplicate per CLSI EP06-A, with polynomial regression modeling.

2.4 Interference Testing and Statistical Analysis

RF interference susceptibility was investigated by: 1) screening 68 RF IgM+ samples for anti-$\alpha v \beta 6$ IgG reactivity, and 2) spiking high-titer RF (>3000 U/mL) into three anti-$\alpha v \beta 6$ IgG-positive samples. Clinical specificity was calculated per CLSI EP12-A2. Statistical analyses employed SPSS 19.0. Categorical data were expressed as percentages with χ^2 tests for group comparisons. Precision CVs < 10% were deemed acceptable. Normality was assessed for continuous variables; non-parametric tests (Mann-Whitney U) were applied to skewed distributions. Significance was defined as $p < 0.05$ ($\alpha = 0.05$).

Table 1. Anti-RF interference ability.

Samples	Control (AU/mL)	ADD RF samples(AU/mL)	Bias
Sample1	10.76	9.92	−7.81%
Sample2	20.98	20.3	−3.24%
Sample3	326.38	316.97	−2.88%

Table 2. Clinical specificity of anti-αvβ6 IgG assay

Population	Age	N	Positive Number	Specificity
Children	6~18(11.6)	128	0	100%
Adult	18~60(36.8)	132	2	99.36%
Elder	60~81(71.2)	100	1	99.00%
Pregnancy	22~37(28.4)	100	0	100%
Total	6~81(32.7)	460	3	99.35%

3 Result

3.1 Diagnostic Performance and Cut-Off Validation

ROC analysis of serum samples from 186 ulcerative colitis (UC) patients and 244 healthy controls established the optimal diagnostic threshold for anti-αvβ6 IgG at 20 AU/mL, maximizing the Youden index. This cutoff yielded a sensitivity of 72.7% and specificity of 94.7%, with an AUC of 0.846 (95% CI) (Fig. 1). Clinical specificity was further validated in 531 individuals across key demographic strata: pediatric (<18 years), adult (18–60 years), elderly (>60 years), and pregnant cohorts (20–43 years). The assay demonstrated exceptional specificity: 100% in pediatric and pregnant groups, 99.36% in adults, and 99.00% in the elderly, resulting in an aggregate specificity of 99.44% (Table 2).

3.2 Analytical Robustness

Precision studies conducted per CLSI EP05-A3 revealed outstanding repeatability (within-run CV: 2.62%-2.99%) and within-laboratory precision (total CV: 5.08%–6.08%) across negative, positive, and high-positive sample controls. Linearity assessment across the analytical range (3-4000 AU/mL) confirmed optimal fit to a first-order polynomial regression model ($Y = a + bX$, $r = 0.99$; Fig. 2), supporting accurate quantification throughout the dynamic range.

3.3 Interference Resistance and Clinical Utility

The assay exhibited complete resistance to rheumatoid factor (RF) interference: all 68 RF IgM-positive samples tested negative for anti-αvβ6 IgG. Crucially, spiking experiments with high-titer RF (>3000 U/mL) into three anti-αvβ6 IgG-positive samples showed no significant deviation from baseline measurements (recovery 98.2–101.7%; Table 1). Combined with the demographic-specific high clinical specificity, these data validate the assay's reliability for clinical application in diverse populations and confirm anti-αvβ6 IgG as a robust serological biomarker for UC diagnosis in the Chinese cohort.

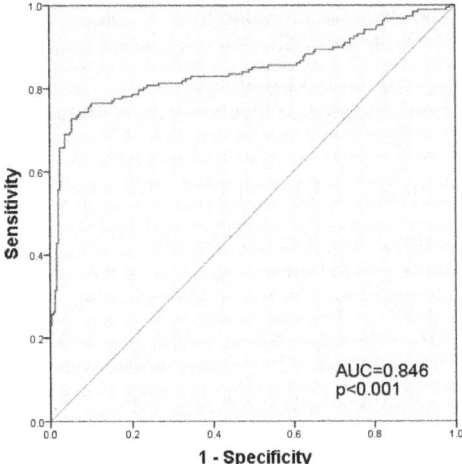

Fig. 1. The area under the receiver operating characteristic curve based on the detections of anti-αvβ6 IgG in healthy individuals and UC patients.

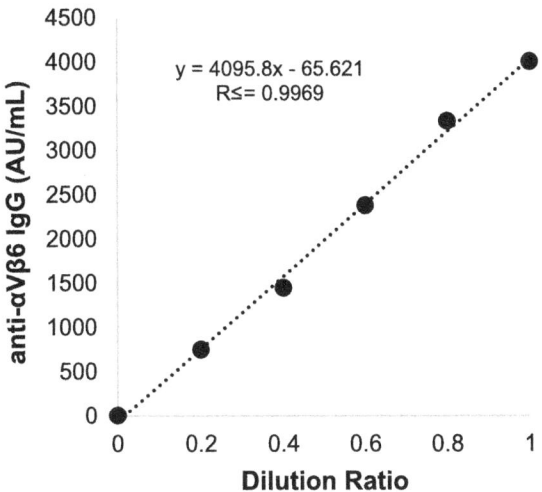

Fig. 2. Calibration curve linearity for anti-αvβ6 IgG assay.

4 Discussion

4.1 Pathogenic Significance of αvβ6 Dysregulation

Integrin αvβ6 emerges as a biologically significant mediator in ulcerative colitis pathogenesis, demonstrating near-undetectable expression within healthy intestinal epithelium yet undergoing substantial induction during active mucosal inflammation. This differential expression profile, quantifiably established through immunohistochemical analyses that correlate staining intensity with endoscopic disease activity, provides a compelling molecular rationale for targeting this receptor in diagnostic strategies. The mechanistic relevance is further reinforced by the functional impact of anti-$\alpha v \beta 6$ IgG autoantibodies, which disrupt critical $\alpha v \beta 6$-fibronectin interactions and consequently impair mucosal restitution processes in ulcerative colitis patients.

4.2 Diagnostic Performance and Transethnic Validation

Our validation of a chemiluminescent immunoassay for anti-$\alpha v \beta 6$ IgG detection within a Southern Chinese cohort yielded clinically robust parameters, including 72.7% sensitivity and 99.44% aggregate specificity across pediatric, adult, geriatric, and pregnant subpopulations. These findings exhibit substantial concordance with Japanese and Swedish reports documenting 92.0% and 76.3% sensitivity respectively, while observed discrepancies with Italian cohorts demonstrating 51.8% positivity likely reflect methodological variations in threshold determination, heterogeneity in disease activity states, or population-specific immunogenetic determinants that necessitate standardized multiethnic evaluation frameworks.

4.3 Comparative Clinical Utility and Implementation Potential

Serological assessment of anti-$\alpha v \beta 6$ IgG confers distinct advantages over conventional ulcerative colitis diagnostic modalities, principally through its non-invasive specimen collection that enhances patient adherence compared to endoscopic procedures. The methodology additionally offers superior analytical specificity relative to fecal biomarkers such as calprotectin and outperforms serum lactoferrin testing in discriminating ulcerative colitis from non-inflammatory bowel conditions. These attributes collectively position $\alpha v \beta 6$ autoantibody quantification as a clinically valuable adjunctive tool capable of facilitating early disease detection and enabling longitudinal therapeutic monitoring.

4.4 Methodological Constraints and Future Investigative Directions

Several study limitations warrant deliberate consideration, foremost being the geographically restricted cohort derivation that constrains broader extrapolation of established diagnostic thresholds across China's heterogeneous populations. The absence of paired endoscopic scoring data precludes assessment of biomarker-disease activity correlations, while fundamental questions regarding antibody kinetics during disease progression remain unresolved. Future investigations should prioritize multi-regional validation cohorts, systematic correlation of serological titers with endoscopic severity indices, combinatorial diagnostic algorithm development, mechanistic exploration of autoantibody ontogeny, and therapeutic assessment of $\alpha v \beta 6$ pathway modulation.

We appreciate the reviewer's insightful suggestion to conduct direct comparisons with established biomarkers such as pANCA, ASCA, and fecal calprotectin to precisely define the incremental diagnostic value of anti-$\alpha v \beta 6$ IgG within clinical algorithms. While direct head-to-head comparisons within our current cohort were not feasible within the scope of this initial study, we have supplemented our discussion with comparative analyses based on existing literature [17, 18]. These analyses highlight potential differences in performance characteristics and suggest a promising role for ant-$\alpha v \beta 6$-IgG. We fully acknowledge the critical importance of empirical, direct comparisons for robust clinical algorithm development. Therefore, a key priority for future research involves specifically designed validation studies within prospective cohorts where all relevant biomarkers, including anti-αvβ6 IgG, pANCA, ASCA, and fecal calprotectin, will be measured simultaneously. This will allow a rigorous assessment of the standalone and combined (incremental) diagnostic and prognostic utility of anti-αvβ6 IgG relative to current standards.

While this study successfully establishes the diagnostic accuracy of anti-αvβ6 IgG detection in identifying UC within our cohort, the relationship between antibody titers and disease activity represents a critical avenue for future investigation. As our primary objective was focused on validating the biomarker's diagnostic performance (sensitivity and specificity), correlating anti-αvβ6 IgG levels with established indices of clinical disease activity – such as the Mayo endoscopic subscore, partial Mayo score, or objective markers of inflammation – fell outside the immediate scope of this foundational work. Nevertheless, understanding this potential relationship is of significant clinical importance. Determining whether anti-αvβ6 IgG levels fluctuate with disease state (e.g., remission vs. flare), correlate with endoscopic severity, or predict treatment response or relapse risk, could substantially enhance its utility beyond diagnosis into monitoring and personalized management strategies. Therefore, we explicitly highlight the investigation of the correlation between anti-αvβ6 IgG titers and UC disease activity as a key priority for subsequent research. Future prospective, longitudinal studies specifically designed to measure serial antibody levels alongside comprehensive clinical activity assessments are essential to elucidate the biomarker's potential role in tracking disease course and informing therapeutic decisions.

5 Conclusion

This study establishes a novel chemiluminescent immunoassay (CLIA) utilizing magnetic microsphere technology for the detection of anti-αvβ6 IgG antibodies. Our methodology represents the first comprehensive validation of this diagnostic approach within China, demonstrating robust clinical sensitivity (72.7%) and exceptional specificity (99.44%) across diverse demographic strata. While further multi-center investigations are warranted to confirm broader applicability, the present findings substantiate the utility of this assay as a viable auxiliary tool for ulcerative colitis diagnosis. The methodologically rigorous dataset generated herein provides foundational evidence for future translational studies exploring αvβ6 serology in inflammatory bowel disease pathogenesis and clinical management.

Acknowledgments. I sincerely thank Dr. Zhisheng Dang for her expert guidance in experimental design, and acknowledge the technical support provided by the Central Laboratory of YHLO.

Disclosure of Interests. The authors declare that they have no competing interests.

Supplementary Statement: We appreciate the reviewer's valuable suggestion regarding the inclusion of participants from Northern and Western China to validate the diagnostic thresholds across China's geographically heterogeneous population. We fully acknowledge the importance of such broader validation. However, the primary objective of the current phase of our study was to establish the foundational diagnostic performance characteristics within the specifically recruited Shenzhen cohort. While our findings demonstrate robust performance within this cohort, we recognize that geographical variations could potentially influence diagnostic parameters. Consequently, confirming the universal applicability of these thresholds across diverse regions within China is a critical next step. Future work will involve planned multi-center studies specifically designed to address this question of generalizability.

References

1. Lin, Z., Gan, M., Wang, X., Su, Z.: Burden of uterine cancer in China from 1990 to 2021 and 15-year projection: a systematic analysis and comparison with global levels. Reprod. Health (1), 144 (2024)
2. Xu, L., He, B., Sun, Y., Li, J., Shen, P., Hu, L., et al.: Incidence of inflammatory bowel disease in urban china: a nationwide population-based study. Clin. Gastroenterol. Hepatol. (13), 3379–86 e29 (2023)
3. Gros, B., Kaplan, G.G.: Ulcerative colitis in adults: a review. JAMA **30**(10), 951–965 (2023)
4. Kucharzik, T., Koletzko, S., Kannengiesser, K., Dignass, A.: Ulcerative colitis-diagnostic and therapeutic algorithms. Dtsch. Arztebl. Int. **117**(33–34), 564–574 (2020)
5. Iwasa, R., Yamada, A., Sono, K., Furukawa, R., Takeuchi, K., Suzuki, Y.: C-reactive protein level at 2 weeks following initiation of infliximab induction therapy predicts outcomes in patients with ulcerative colitis: a 3 year follow-up study. BMC Gastroenterol. **15**, 103 (2015)
6. Ruemmele, F.M., Targan, S.R., Levy, G., Dubinsky, M., Braun, J., Seidman, E.G.: Diagnostic accuracy of serological assays in pediatric inflammatory bowel disease. Gastroenterology **115**(4), 822–829 (1998)
7. Rump, J.A., Scholmerich, J., Gross, V., Roth, M., Helfesrieder, R., Rautmann, A., et al.: A new type of perinuclear anti-neutrophil cytoplasmic antibody (p-ANCA) in active ulcerative colitis but not in Crohn's disease. Immunobiology **181**(4–5), 406–13 (990)
8. Vasiliauskas, E.A., Plevy, S.E., Landers, C.J., Binder, S.W., Ferguson, D.M., Yang, H., et al.: Perinuclear antineutrophil cytoplasmic antibodies in patients with Crohn's disease define a clinical subgroup. Gastroenterology **110**(6), 1810–1819 (1996)
9. Plevy, S., Silverberg, M.S., Lockton, S., Stockfisch, T., Croner, L., Stachelski, J., et al.: Combined serological, genetic, and inflammatory markers differentiate non-IBD, Crohn's disease, and ulcerative colitis patients. Inflamm. Bowel Dis. **19**(6), 1139–1148 (2013)
10. Kadry, Y.A., Calderwood, D.A.: Chapter 22: structural and signaling functions of integrins. Biochim. Biophys. Acta Biomembr. **1862**(5), 183206 (2020)
11. Humphries, J.D., Byron, A., Humphries, M.J.: Integrin ligands at a glance. J. Cell Sci. **119**(19), 3901–3903 (2006)
12. Sakurai, T., Saruta, M.: Positioning and usefulness of biomarkers in inflammatory bowel disease. Digestion **104**(1), 30–41 (2023)
13. Brzozowska, E., Deshmukh, S.: Integrin Alpha v Beta 6 (alphavbeta6) and its implications in cancer treatment. Int. J. Mol. Sci. **23**(20) (2022)
14. Campbell, I.D., Humphries, M.J.: Integrin structure, activation, and interactions. Cold Spring Harb. Perspect. Biol. **3**(3) (2011)
15. Kuwada, T., Shiokawa, M., Kodama, Y., Ota, S., Kakiuchi, N., Nannya, Y., et al.: Identification of an anti-integrin alphavbeta6 autoantibody in patients with ulcerative colitis. Gastroenterology **160**(7), 2383–94 e21 (2021)

16. Muramoto, Y., Nihira, H., Shiokawa, M., Izawa, K., Hiejima, E., Seno, H., et al.: Anti-Integrin alphavbeta6 antibody as a diagnostic marker for pediatric patients with ulcerative colitis. Gastroenterology **163**(4), 1094–7 e14 (2022)
17. Mokrowiecka, A., Gasiorowska, A., Malecka-Panas, E.: PANCA and ASCA in the diagnosis of different subtypes of inflammatory bowel disease. Hepatogastroenterology **54**(77), 1443–1448 (2007)
18. Fengming, Y., Jianbing, W.: Biomarkers of inflammatory bowel disease. Dis. Markers **2014**, 710915 (2014)

Does Commercial Pension Insurance Participation Promote the Transfer of Land Contractual Management Rights?—An Empirical Study Based on CHARLS

Xinlong Yang[1], Lei Yu[2(✉)], and Xinyi Luo[1]

[1] Shenzhen Institute of Information Technology, Shenzhen, China
[2] Xi'an University of Technology, Xi'an, China
chsdcy@xaut.edu.cn

Abstract. This study investigates the impact of commercial pension insurance participation on rural households' land contractual management rights transfer behavior, using data from the China Health and Retirement Longitudinal Study (CHARLS) in 2018. First, we integrate theories of household decision-making and land transfer to analyze the mechanism through which insurance participation affects land transfer decisions and propose corresponding hypotheses. Second, to address endogeneity, we employ the IV Probit model with the "provincial participation rate" as an instrumental variable, and examine heterogeneous effects across age groups and regions. Robustness checks are conducted by replacing the key explanatory variable (household participation rate) and adjusting the sample period (CHARLS 2015 data). Results show that commercial pension insurance participation significantly inhibits land transfer activities, with heterogeneous effects across age and regional samples.

Keywords: Commercial pension insurance participation · Land renting out · Land renting in

1 Introduction

1.1 Research Background and Significance

As the third pillar of China's pension system, commercial pension insurance addresses gaps in basic pension coverage (first pillar) and enterprise annuity accessibility (second pillar) through market-driven savings mechanisms. Despite low rural penetration, it enhances elderly care quality and social stability by stabilizing income expectations and curbing irrational consumption among seniors. Rural elderly support currently relies on informal mechanisms—primarily "family support" and "land-based security." Land functions dualistically: as a social safety net for returning migrant workers and an economic asset generating income via land-use rights. This "land binding" effect results in underutilized farmland, with nationwide transfer rates covering only one-third of total arable area, while smallholder farming persists as the dominant model. Policy efforts, such as the Opinions on Comprehensively Promoting Rural Revitalization and Accelerating Agricultural and Rural Modernization, explicitly advocate for improved farmland transfer systems to enable moderate-scale operations. Yet, practical reforms to relax rural-land constraints face persistent challenges. Within this context, commercial pension insurance may influence farmers' land decisions by reshaping elderly care expectations: while augmented income security could incentivize land transfers, strengthened risk protection might deter such transfers. The net effect remains unclear, as does its potential variation across age groups and regions—key questions requiring empirical validation.

While the question of whether institutional pensions can fully substitute land - based pensions remains unresolved, pensions play a significant role in promoting land transfer and rural revitalization. On one hand, they facilitate the urbanization of agricultural transfer populations and accelerate land circulation. Migrant workers often adopt "part - time farming" or "inter - generational relay" models to cope with financial pressures, constraining farmland transfer despite economic rationality, which hinders agricultural modernization. Future reforms (e.g., homestead transfer) should reduce urbanization costs to promote whole - family migration and release land resources. On the other hand, pensions stabilize long - term income expectations, enabling farmers to reduce land dependence and transition to non - agricultural employment, thereby optimizing resource allocation from inefficient to efficient farmers through the "separation of three rights" reform.

1.2 Literature Review

Research on Factors Influencing Farmland Transfer. Scholars have examined farmland transfer drivers across macro and micro levels. Macro-level factors include market standardization, policy efficiency, and financial development: international evidence highlights China's gaps in policy support and market imperfections, while the US "free migration model" and Japan's "leapfrog transfer + non-agricultural shift" show transfers occur when gains exceed farming income (Li 2022). Zhu et al. (2021) attribute China's smaller transfer scale to market deficiencies and contract gaps. Financial development boosts transfer willingness via credit access and labor reallocation (Zhang & Lu 2022), with policy awareness shaping decisions through psychological expectations (Zhang et al. 2023). Land titling reforms enhance motivation by enabling land function transformation (Li 2022). Regional disparities persist: elderly farmers in eastern regions inhibit transfers, while central provinces facilitate them (Zhang 2017), and digital finance reduces transaction costs (Zhang 2022). Micro-level factors involve household income structures, labor mobility, and elder-care modes: agricultural income reduces transfers but increases land acquisitions, with married households favoring acquisitions (Yang & Peng 2021; Mao et al. 2020). Labor mobility suppresses acquisitions and promotes transfers, as stable employment (e.g., spouse-to-children migration) creates "part-time farming" barriers (Miao et al. 2021), while unstable jobs deter elderly, tradition-bound males (Xu et al. 2022). Family care (grandchild-care boosts transfers, parental-care inhibits them) and social services further shape outcomes (Wu et al. 2020; Zhang &

Zhuang 2022). Ultimately, transfers reflect the interplay of economic income, livelihood strategies, and family structures (Wang, 2022).

Literature on the Impact of Social Security on Farmland Transfer. The impact of Social security influences farmland transfer primarily through substitution effects. Zhu and Wu (2022) established an evolutionary game model involving farmers, agribusinesses, and insurers, demonstrating that equilibrium—shaped by government subsidies, firm efficiency, and insurance pricing—optimizes transfer insurance schemes. Pension participation substitutes land rights, with exogenous land entitlements and pension investments negatively correlated, reducing elderly care risks and economic uncertainties via transfer payments (Yu et al. 2010; Xu & Zhou, 2019). Institutional security, exemplified by China's New Rural Pension Scheme (NRPS), replaces land's social safety net, altering rental price expectations and curbing inflationary pressures (Li et al. 2019). Empirically, Zhang et al. (2019) found NRPS increased transfers among farmers aged ≥ 60. Xu and Zhou (2019) noted stronger substitution effects for households without elderly members or with weak liquidity constraints, while Luo et al. (2019) observed NRPS promoted transfers out but not in. Contrarily, Mao et al. (2020) argued pension insurance capitalizes farmland, boosting bidirectional transfers, whereas Wu et al. (2020) contended NRPS-induced income insufficiently replaces farmland returns, potentially inhibiting transfers due to urban-rural disparities. Commercial pension insurance indirectly facilitates transfers by improving elderly life satisfaction (Sun et al. 2019). Zheng and Wang (2022) showed it reduces farmland inputs and enhances leisure time, promoting transfers—particularly in western China—though scale effects vary regionally.

Review of Existing Literature. Existing studies demonstrate that although numerous factors influence farmland transfer, these factors are not independent of each other and exhibit varying degrees of impact, including macro-level factors such as financial development and micro-level factors such as household income structure. Regarding the impact of social security on farmland transfer, the prevailing view is that institutional security exhibits a substitution effect with farmland transfer, and significant progress has been made in this research area. However, two major gaps remain: first, while existing studies predominantly focus on the impact of the New Rural Pension Scheme (NRPS) on farmland transfer, there is limited literature examining the role of commercial pension insurance—the third pillar of social security. Second, empirical research on the influence of commercial pension insurance participation on land contracting rights transfer remains scarce. Although Zheng and Wang (2022) provide empirical evidence that commercial pension insurance participation and income levels significantly affect farmland transfer, their study does not specifically focus on land contracting rights transfer. This paper aims to address this gap by systematically investigating the impact mechanism of commercial pension insurance participation on land contracting rights transfer.

2 Research Hypothesis

Grounded in farmer behavior theory, Scott's "subsistence ethic" posits that farmers prioritize survival security over income maximization, only leasing out land contracting rights when third-pillar insurance coverage is adequately established (Xu & Zhou 2019).

Farmers opt to rent land when combined non-farm wages and land rents exceed agricultural income; they choose to rent in land when agricultural marginal returns surpass opportunity costs and rental expenses, achieving Pareto optimality when these margins equalize. Existing research confirms that pension insurance substitutes land security functions, significantly promoting land transfers (Mao Fei et al. 2020).

Based on this, we propose:

Hypothesis 1a: Commercial pension insurance participation positively affects land rental behavior.

Hypothesis 1b: Commercial pension insurance participation positively affects land leasing behavior.

3 Research Design

3.1 Data Sources

This study utilizes the 2018 cross-sectional data from the China Health and Retirement Longitudinal Study (CHARLS), which covers household and community-level information, incorporating key variables such as land rental arrangements. By matching 13 sub-databases through household ID and community ID, we retained 18,036 valid observations after excluding missing values, with proportionally balanced regional distribution (33.38% East, 33.25% Central, 33.37% West).

For empirical analysis, we employ regression models and Probit marginal effects to quantify the impact of commercial pension insurance participation. Instrumental variable approaches are applied to address potential endogeneity, while subgroup and regional analyses are conducted to examine heterogeneous effects and ensure robustness.

3.2 Variable Selection

Explanatory Variables. Given that farmland transfer decisions are typically made at the household level, this study selects a household-level indicator to more accurately measure commercial pension insurance participation. The CHARLS database provides detailed records of each family member's commercial pension insurance status. Drawing on prior research, we construct a binary dummy variable indicating whether any household member participates in commercial pension insurance (1 = yes, 0 = no) as the core explanatory variable.

Dependent Variables. Existing literature predominantly operationalizes farmland transfer behavior as a binary dummy variable, with some studies examining transfer scale or rental payments (Zheng & Wang 2022). Focusing on the impact of commercial pension insurance participation on land contracting rights transfer, this study defines the dependent variable based on the survey question "Did you rent out any land to others in the past year?" Responses are coded as 1 ("Yes") for land rental and 0 ("No") otherwise. Drawing on prior research (Mao Fei et al. 2020; Zhao & Li 2014), we additionally incorporate land leasing behavior as a dependent variable, operationalized through the survey question "Did you rent any land in the past year?" with responses coded as 1 ("Yes") for leasing and 0 ("No") otherwise. This dual-variable approach enables comprehensive examination of the relationship between insurance participation and land transfer decisions.

Control Variables. To accurately measure the net effect of commercial pension insurance participation on land contracting rights transfer and enhance result reliability, this study selects control variables from individual, family, and regional levels based on research questions and CHARLS data.

At the individual level, although land transfer decisions are typically made at the household level, household heads play a pivotal role. This study treats CHARLS respondents as proxy household heads and incorporates demographic variables including gender, age, marital status, education level, and health status. Following the life-cycle hypothesis, we include age squared to capture potential nonlinear effects.

For family-level variables, we select family size, per capita net income, agricultural labor ratio, and farm machinery value. Per capita net income reflects the impact of household economic conditions on pension security and land dependence; agricultural labor ratio measures household agricultural production capacity; farm machinery value indirectly affects land transfer decisions by influencing production efficiency (Leng et al. 2015).

Regional variables include land rental rates and provincial dummy variables. Land rental rates affect tenants' willingness and transfer efficiency (Zhang & Wan 2007; Kong & Xu 2010). Provincial dummies control for regional economic disparities in land transfer market development (Mao Fei et al. 2020; Zhao & Li 2014).

3.3 Descriptive Statistics

The definitions of specific variables and their descriptive statistics are presented in Table 1.

Table 1. Variable Definitions and Descriptive Statistical Analysis.

Variable types	Variable	Variable Description	Mean.	Std. Dev
Dependent variable	Land renting out (yes/no)	Whether the family has rented out the land (1 = yes; 0 = no)	0.16	0.36
	Land renting in (yes/no)	Whether the family has rented in the land (1 = yes; 0 = no)	0.07	0.25
Independent variable	Insurance participation	Whether family members participate in commercial pension insurance (1 = yes; 0 = no) (Ins_P)	0.23	0.42
Individual characteristics	Gender	Gender of the respondent (1 = Male; 0 = Female)	0.47	0.50

(*continued*)

Table 1. (*continued*)

Variable types	Variable	Variable Description	Mean.	Std. Dev
	Age	Respondent's age (in years, completed birthdays)	66.37	9.88
	Age^2	Squared age of respondent at survey time (in completed years)	4502.25	1350.24
	Married	Marital status (1 = Married; 0 = Separated/Divorced/Widowed/Never married)	0.86	0.34
	Edu	Highest level of education (1 = Junior high school or below; 2 = Senior high school/technical secondary school; 3 = College degree or above)	1.15	0.41
	Health status	Self-rated health status (1 = Excellent; 2 = Very good; 3 = Good; 4 = Fair; 5 = Poor)	2.94	1.02
Household characteristics	household size	Total household size(HS)	1.31	0.4
	Per capita net income of households	Ln(household per capita net income in yuan, 2017–2018) Ln(income)	4.40	4.95
	Proportion of agricultural population in the household	Proportion of household members engaged in agricultural activities (AgriPop.%)	0.85	0.96
	Value of agricultural machinery	Ln(value of household-owned agricultural fixed assets in yuan) Ln(value)	1.43	2.81
Regional characteristics	Rental level	Ln(land transfer rental income in yuan)	0.95	2.43
	Region	Region (1 = East; 2 = Central; 3 = West)	2.00	0.82

Table 2 indicates an average commercial pension insurance participation rate of 23.49% in the total sample, reflecting insufficient awareness of self-funded retirement planning among Chinese residents. Regional analysis shows participation rates of 26.13%, 25.11%, and 19.22% in the eastern, central, and western regions respectively, with the eastern and central regions significantly outpacing the western region.

Table 3 shows that the average participation rates in land leasing behaviors are 15.57% and 6.97% respectively, indicating a generally underdeveloped farmland transfer market in China where farmers show limited willingness to transfer land use rights, particularly with significantly lower leasing willingness than rental activity. Regional

Table 2. Commercial pension insurance participation across regions

	Total	East	Central	West
Participation in Commercial Pension Insurance	4236	1573	1506	1157
Non-participation in commercial pension insurance	13800	4447	4491	4862
Prob. of commercial pension participation	23.49%	26.13%	25.11%	19.22%

analysis reveals participation rates of 16.23%, 16.81%, and 13.69% for land rental in the eastern, central, and western regions respectively, with the eastern and central regions significantly outpacing the western region; leasing participation rates stand at 6.03%, 8.60%, and 6.28% across the same regions, where the central region exhibits significantly higher rates than both the eastern and western regions.

Table 3. Land transfer situations across regions

	Total	East	Central	West
Land renting out	2809	977	1008	824
Non-land renting out	15227	5043	4989	5195
Prob. of land renting out behavior	15.57%	16.23%	16.81%	13.69%
Land renting in	1257	363	516	378
Non-land renting in	16779	5657	5481	5641
Prob. of land renting in behavior	6.97%	6.03%	8.60%	6.28%

3.4 Model Specification

Probit Model. To examine the binary decision - making of land leasing, this study employs the Probit model as the baseline.

$$P(rent_i = 1) = \Phi(a_0 + a_1 insurance_i + \beta X_i + \varepsilon_i)$$

Let $rent_i$ denote the land transfer behavior of the i - the household (further specified into $rentout_i$ and $rentin_i$ in empirical analysis), $insurance_i$ represent the commercial pension insurance participation status, X_i be the vector of control variables including individual, family, and regional characteristics, and ε_i the random error term.

4 Empirical Analysis

4.1 Benchmark Regression Results

To examine the impact of commercial pension insurance on land leasing behaviors, this study employs a stepwise regression approach to test model stability and variable effects by sequentially introducing control variables. Given that Probit coefficients lack intuitive economic interpretation, we report both OLS coefficients and average marginal effects.

The results indicate that the "insurance participation" variable is significantly negative at the 1% level, with marginal effects of −0.091, −0.108, and −0.014 (all significant at 1%), implying that insured households are 9.1%, 10.8%, and 1.4% less likely to rent out land than uninsured households, respectively. This contradicts Hypothesis 1a, suggesting that commercial pension insurance suppresses land rental willingness, possibly because its income cannot fully replace farming income (Wu Wanpei et al. 2020).

The analysis of control variables yields three key insights. At the individual level, age exhibits a U-shaped relationship with land rental behavior, with younger adults demonstrating the highest probability of participation (Luan & Ma 2021), while gender, marital status, and health status show no statistically significant effects. Regarding family characteristics, larger household size and higher agricultural labor ratios significantly decrease the likelihood of land rental, suggesting that households with abundant labor prefer off-farm employment while retaining land. Per capita income and farm machinery value do not significantly influence rental decisions. At the regional level, higher land rental prices are associated with increased rental probability, reflecting an "economic account" effect (Wang 2022). Notably, regional dummies indicate significantly higher rental probabilities in eastern China, where economic development reduces dependence on land security.

Table 4. Empirical Estimates of Commercial Pension Insurance's Effect on Land Renting out Behavior

Variable	Land renting out (yes/no)					
	OLS			Probit Marginal Effects		
	(1)	(2)	(3)	(4)	(5)	(6)
Insured or not	−0.081***	−0.102***	−0.017***	−0.091***	−0.108***	−0.014***
	(−13.06)	(−15.16)	(−5.47)	(−11.86)	(−13.60)	(−4.19)
Gender	0.009	0.010*	0.001	0.009*	0.011**	0.001
	(1.56)	(1.91)	(0.41)	(1.74)	(2.02)	(0.23)
Age	0.008***	0.010***	−0.004**	0.009***	0.011***	−0.003***
	(3.06)	(3.58)	(-2.57)	(3.00)	(3.47)	(−2.94)
Age2	−0.000***	−0.000***	0.000**	−0.000***		0.000***
	(−3.22)	(−3.69)	(2.45)	(−3.14)		(2.83)
Married	0.006	0.015*	−0.009**	0.007	0.014*	−0.004
	(0.76)	(1.73)	(-2.14)	(0.82)	(1.65)	(−1.16)
Edu	−0.017***	−0.027***	−0.006**	−0.024***	−0.032***	−0.004
	(−2.87)	(−4.39)	(−2.11)	(−3.05)	(−4.09)	(−1.43)
Health statu	−0.007**	−0.005*	−0.001	−0.006**	−0.005*	−0.001
	(−2.39)	(−1.82)	(−0.56)	(−2.41)	(−1.77)	(−0.92)
Household size		−0.028***	−0.006***		−0.029***	−0.006***

(continued)

Table 4. (*continued*)

Variable	Land renting out (yes/no)					
	OLS			Probit Marginal Effects		
	(1)	(2)	(3)	(4)	(5)	(6)
		(−7.09)	(−3.38)		(−6.35)	(−3.29)
Ln(income)		0.003***	0.000		0.003***	0.000
		(5.61)	(1.13)		(5.61)	(0.50)
Agri pop.%		−0.032***	−0.007***		−0.030***	−0.008***
		(−9.71)	(−5.54)		(−9.17)	(−5.83)
Ln(value)		−0.000	−0.002***		−0.000	−0.001
		(−0.10)	(−4.08)		(−0.09)	(−1.43)
Rental level			0.135***			0.042***
			(270.53)			(15.51)
Region			0.002			0.004***
			(1.15)			(2.80)
Sample size	18036	18036	18036	18036	18036	18036

Note: Standard deviations are in parentheses and ***, **, and * indicate significant at the 1%, 5%, and 10% levels, respectively. The marginal effects estimated by the Probit model are reported in the table. The same as below.

The empirical results in Table 5 indicate that the regression coefficient of the "insurance participation" variable is significantly negative at the 1% level, with marginal effects of −0.059, −0.030, and −0.031 (all significant at the 1% level), suggesting that insured households are 5.9%, 3.0%, and 3.1% less likely to rent out land than uninsured households, respectively. This demonstrates that commercial pension insurance participation inhibits land rental willingness, contradicting Hypothesis 1b. The underlying reason may lie in farmers' limited and unstable income sources; participation in commercial pension insurance reduces personal savings, thereby weakening their enthusiasm for agricultural land rental.

Analysis of control variables reveals: (1) Individual characteristics: gender positively affects land rental; age shows an inverted U-shaped relationship (peak in young adulthood, Chen & Zhai 2015; Luan & Ma 2021). Married households rent more (labor advantage, Mao et al. 2020), while better health reduces rental likelihood (preference for non-farm jobs, Mao et al. 2020). (2) Family characteristics: Larger family size and higher agricultural labor ratio increase rentals (labor abundance). Per capita income negatively correlates with rentals (economic security effect, Wang 2022). Higher farm machinery value boosts rentals (mechanization demand). (3) Regional characteristics: Higher rental prices decrease rental probability (scale farming disincentive, Wang 2022). Western region dummies show negative effects (migrant land retention).

Table 5. Empirical Estimates of Commercial Pension Insurance's Effect on Land Renting in Behavior

Variable	Land renting in (yes/no)					
	OLS			Probit Marginal Effects		
	(1)	(2)	(3)	(4)	(5)	(6)
Insured or not	−0.048***	−0.013***	−0.016***	−0.059***	−0.030***	−0.031***
	(−12.05)	(−3.38)	(−4.07)	(−10.14)	(−5.24)	(−5.54)
Gender	0.013***	0.007*	0.008**	0.013***	0.009**	0.009***
	(3.3)	(1.86)	(1.99)	(3.49)	(2.50)	(2.60)
Age	−0.004***	−0.007***	−0.007***	0.011***	0.007**	0.007**
	(−1.81)	(−3.59)	(−3.56)	(3.51)	(2.37)	(2.26)
Age2	0.000	0.000**	0.000**	−0.000***	−0.000***	−0.000***
	(0.33)	(2.41)	(2.35)	(−4.52)	(−3.31)	(−3.23)
Married	0.037***	0.010***	0.010***	0.061***	0.027***	0.026***
	(10.60)	(2.81)	(2.78)	(7.23)	(3.35)	(3.28)
Edu	−0.020***	−0.004	−0.004	−0.026***	−0.009	−0.009
	(−5.15)	(−1.12)	(−1.19)	(−4.43)	(−1.60)	(−1.56)
Health statu	−0.004**	−0.005***	−0.004**	−0.004*	−0.004**	−0.003*
	(−1.99)	(−2.84)	(−2.35)	(−1.95)	(−2.44)	(−1.90)
Household size		0.007**	0.006**		0.012***	0.012***
		(2.24)	(2.19)		(4.49)	(4.49)
Ln(income)		−0.002***	−0.002***		−0.001***	−0.001***
		(−3.57)	(−3.86)		(−2.93)	(−3.29)
Agri pop.%		0.040***	0.040***		0.037***	0.037***
		(17..25)	(17.23)		(18.80)	(18.74)
Ln(value)		0.013***	0.014***		0.008***	0.008***
		(13.31)	(13.56)		(14.55)	(14.75)
Rental level			−0.004***			−0.004***
			(−6.17)			(−4.59)
Region			−0.011***			−0.010***
			(−4.86)			(−4.35)
Sample size	18036	18036	18036	18036	18036	18036

Note: Standard deviations are in parentheses and ***, **, and * indicate significant at the 1%, 5%, and 10% levels, respectively. The marginal effects estimated by the Probit model are reported in the table. The same as below

4.2 Endogeneity Treatment

To address potential endogeneity from bidirectional causality and omitted variables, we employ the Instrumental Variable Probit model and select the historical provincial commercial pension insurance participation rate as the instrument variable, constructed by matching 2021 provincial rates with households' provincial affiliations from the micro-survey.

The instrument satisfies both relevance and exogeneity criteria. Relevance is confirmed by its direct influence on current insurance decisions through historical participation patterns. Exogeneity holds as provincial policies determine the rate, ensuring no direct correlation with household or regional land transfer decisions.

Table 6 presents first-stage results and endogeneity tests. Column (1) shows the provincial rate's marginal effect on insurance participation is significantly negative at 1%. Endogeneity tests reject the null hypothesis ($p < 0.1$), while weak instrument tests (AR/Wald, $p < 0.1$) confirm strong first-stage correlations. The just-identified model precludes over-identification tests, with all instrument variables meeting exogeneity requirements.

Table 6. Endogeneity Treatment

	Whether insured	Whether land rented out	Whether land rented in
	(1)	(2)	(3)
Insured or not		−0.013***	−0.031***
		(−4.11)	(−5.41)
Provincial Commercial Pension Insurance Participation Rate	−0.131***	0.020*	0.062***
	(−5.63)	(1.82)	(3.81)
Gender	0.043***	0.000	0.009**
	(7.59)	(0.21)	(2.56)
Age	0.000	−0.003***	0.007**
	(0.13)	(−2.91)	(2.23)
Age2	0.000	0.000***	−0.000***
	(0.37)	(2.82)	(−3.18)
Married	0.057***	−0.004	0.027***
	(6.69)	(−1.13)	(3.33)
Edu	0.246***	−0.005	−0.009*
	(41.07)	(−1.49)	(−1.68)
Health statu	−0.025***	−0.001	−0.004**
	(−9.32)	(−1.00)	(−2.03)
Household size	−0.034***	−0.006***	0.012***
	(−7.54)	(−3.30)	(4.60)

(*continued*)

Table 6. (continued)

	Whether insured	Whether land rented out	Whether land rented in
	(1)	(2)	(3)
Ln(income)	0.005***	0.000	−0.001***
	(7.64)	(0.51)	(−3.37)
Agri pop.%	−0.099***	−0.008***	0.037***
	(−26.46)	(−5.92)	(18.62)
Ln(value)	−0.006***	−0.001	0.008***
	(−4.70)	(−1.52)	(14.59)
Rental level	−0.015***	0.042***	−0.004***
	(−11.50)	(15.68)	(−4.56)
Region	−0.003	0.003*	−0.013***
	(−0.94)	(1.93)	(−5.26)
Sample size	18036	18036	18036
IV t − value		−6.18	−6.18
Initial IV Test			
P-value		0.0361	0.0002
Weak IV Test			
AR P-value		0.0212	0.0001
Wald P-value		0.0290	0.0005

Note: Standard deviations are in parentheses and ***, **, and * indicate significant at the 1%, 5%, and 10% levels, respectively. The marginal effects estimated by the Probit model are reported in the table. The same as below.

4.3 Heterogeneity Analysis

Considering that the increasing proportion of land leasing by new agricultural business entities (e.g., cooperatives) is primarily driven by policy support rather than endogenous factors, this study focuses on "land rental decisions" to examine age and regional heterogeneity.

Age Heterogeneity Test. In the context of rapid industrialization and urbanization, the aging of rural households has intensified while the off-farm employment rate among young laborers has risen. As shown in Table 7, the marginal effects of commercial pension insurance participation are significantly negative at the 1% level for elderly households and 5% level for middle-aged households, but insignificant for transitional households. This confirms the validity of using 65 years as the threshold, highlighting discrepancies between traditional economic age classifications and rural China's realities.

Further analysis suggests two possible mechanisms: (1) an inverted U-shaped relationship between age and land rental willingness, where rental probability declines beyond a certain age threshold (Wu et al. 2016); (2) elderly households without intergenerational labor succession tend to retain land (Le Zhang 2010).

Table 7. Regression Results of Age - based Heterogeneity Analysis

Variable	Land renting out (yes/no)		
	(1) >65 years old	(2) 60–65 years old	(3) <60 years old
Insured or not	−0.018***	−0.011	−0.012**
	(−3.87)	(−1.57)	(−1.98)
Gender	0.000	0.004	−0.001
	(0.03)	(0.81)	(−0.30)
Married	−0.005	−0.015*	−0.002
	(−1.47)	(−1.77)	(−0.21)
Edu	−0.002	−0.002	−0.012**
	(−0.45)	(−0.39)	(−2.05)
Health status	0.001	0.001	−0.005**
	(0.52)	(0.25)	(−2.16)
Household size	−0.005	−0.007	−0.008**
	(−1.44)	(−1.50)	(−2.41)
Ln(income)	−0.002***	0.001	0.002***
	(−3.53)	(1.37)	(2.90)
Agri pop.%	−0.004**	−0.005	−0.024***
	(−2.11)	(−1.51)	(−6.14)
Ln(value)	−0.001	−0.001	0.000
	(−0.92)	(−0.91)	(0.26)
Rental level	0.040***	0.000	0.000
	(15.25)	(.)	(.)
Region	0.005**	0.003	0.003
	(2.54)	(0.86)	(0.83)
Sample size	9347	2918	4591

Note: Standard deviations are in parentheses and ***, **, and * indicate significant at the 1%, 5%, and 10% levels, respectively. The marginal effects estimated by the Probit model are reported in the table. The same as below.

Regional Heterogeneity Test. Significant disparities exist in commercial pension insurance participation and land transfer marketization across eastern, central, and western China. Table 8 shows that the marginal effect of commercial pension insurance participation is significantly negative at the 1% level for both central and western regions, but insignificant for the eastern region. This may stem from the eastern region's developed economy and diverse non - farm employment opportunities, which weaken the relationship between insurance participation and land rental decisions (Zheng & Wang 2022).

Table 8. Regression Results of Regional Heterogeneity

variable	Land renting out (yes/no)		
	(1) East	(2) Central	(3) West
Insured or not	−0.007	−0.023***	−0.017***
	(−1.46)	(−3.59)	(−2.76)
Gender	0.001	0.001	0.000
	(0.24)	(0.14)	(0.07)
Age	−0.004**	−0.005**	0.000
	(−2.06)	(−2.53)	(0.17)
Age2	0.000*	0.000**	0.000
	(1.91)	(2.34)	(0.03)
Married	0.005	−0.007	−0.009*
	(0.79)	(−1.07)	(−1.67)
Edu	−0.008	0.003	−0.009
	(−1.43)	(0.44)	(−1.36)
Health statu	0.000	−0.002	−0.002
	(0.25)	(−0.72)	(−0.91)
Household size	−0.007**	−0.001	−0.013***
	(−2.07)	(−0.39)	(−3.27)
Ln(income)	−0.000	−0.001**	0.002***
	(−0.34)	(−2.24)	(3.45)
Agri pop.%	−0.010***	−0.008***	−0.010***
	(−3.53)	(−2.75)	(−3.94)
Ln(value)	0.001	−0.001	−0.001
	(0.89)	(−1.43)	(−1.24)
Rental level	0.000	0.000	0.043***
	(.)	(.)	(13.81)
Sample size	5145	5117	6019

Note: Standard deviations are in parentheses and ***, **, and * indicate significant at the 1%, 5%, and 10% levels, respectively. The marginal effects estimated by the Probit model are reported in the table. The same as below.

4.4 Robustness Checks

To verify the robustness of benchmark regression results, this study employs two approaches: replacing the explanatory variable and adjusting the sample period.

Replacement of Explanatory Variable. The original variable "whether any family member participates in commercial pension insurance" is replaced with "household insurance participation rate" (ratio of actual insured members to total household size) to more precisely measure intra - household coverage. As shown in Columns (1) and (3) of Tables 4, 5 and 6, the marginal effects of household insurance rate are significantly negative at the 1% level for both land rental and leasing decisions, reducing the probabilities by 1.4 and 2.9 percentage points respectively per unit increase. These findings align with the benchmark regression, confirming that higher family participation rates significantly suppress land transfer willingness.

Adjustment of Sample Period. Using CHARLS2015 data, the regression results in Columns (2) and (4) of Table 9 show that the marginal effects of "insurance participation" remain significant (−0.008 at 5% level for rental, −0.044 at 1% level for leasing), consistent with the benchmark results.

In conclusion, the negative impact of commercial pension insurance on land rental and leasing behaviors is robustly verified.

Table 9. Results of Additional Robustness Tests

variable	Land renting out (yes/no)			Land renting in (yes/no)
	(1) Replace the explanatory var.	(2) Replace with CHARLS2015	(3) Replace the explanatory var.	(4) Replace with CHARLS2015
Prob.	−0.014***		−0.029***	
	(−4.06)		(−4.87)	
Insured or not		−0.008**		−0.044***
		(−2.35)		(−5.66)
Gender	0.000	0.001	0.009***	0.004
	(0.21)	(0.59)	(2.58)	(0.86)
Age	−0.003***	−0.003***	0.007**	0.011***
	(−2.98)	(−2.66)	(2.27)	(3.34)
Age2	0.000***	0.000**	−0.000***	−0.000***
	(2.87)	(2.40)	(−3.24)	(−3.88)
Married	−0.004	−0.009***	0.026***	0.039***
	(−1.17)	(−2.61)	(3.26)	(3.56)
Edu	−0.004	−0.003	−0.009*	−0.007
	(−1.42)	(−1.11)	(−1.68)	(−1.28)

(*continued*)

Table 9. (*continued*)

variable	Land renting out (yes/no)			Land renting in (yes/no)
	(1) Replace the explanatory var.	(2) Replace with CHARLS2015	(3) Replace the explanatory var.	(4) Replace with CHARLS2015
Health statu	−0.001	0.000	−0.003*	−0.003
	(−0.90)	(0.07)	(−1.82)	(−1.21)
Household size	−0.007***	−0.001	0.010***	0.002
	(−3.71)	(−0.69)	(3.97)	(1.24)
Ln(income)	0.000	0.000	−0.001***	−0.003***
	(0.48)	(0.28)	(−3.37)	(−4.39)
Agri pop.%	−0.008***	−0.007	0.037***	0.036***
	(−5.83)	(−1.41)	(18.78)	(5.93)
Ln(value)	−0.001	−0.001***	0.008***	0.010***
	(−1.40)	(−2.94)	(14.79)	(13.59)
Rental level	0.042***	0.025***	−0.004***	−0.002*
	(15.52)	(13.97)	(−4.54)	(−1.77)
Region	0.004***	−0.000	−0.010***	−0.004
	(2.80)	(−0.01)	(−4.34)	(−1.29)
Sample size	18036	9727	18036	9727

Note: Standard deviations are in parentheses and ***, **, and * indicate significant at the 1%, 5%, and 10% levels, respectively. The marginal effects estimated by the Probit model are reported in the table. The same as below.

5 Research Conclusions

Based on CHARLS 2018 dataset, this paper constructs a Probit model to examine how commercial pension insurance participation affects land leasing behavior and explores group differences. Key findings: (1) Participation significantly reduces leasing; (2) Transitional households (aged 60–65) show insignificant effects; (3) Eastern regions exhibit no significant link, likely due to higher economic/agricultural development than central/western areas.

Commercial pension insurance significantly reduces farmers' land rental participation through three mechanisms: First, income substitution replaces land's security function with pension payments, weakening rental income reliance and production expansion needs. Second, risk hedging diminishes motivations to use land transfers for poverty mitigation, causing renters to avoid uncertainty and lessees to shun operational risks, jointly suppressing transfers. Third, asset allocation shifts land from production to non-agricultural uses, cutting short-term rentals. Collectively, these mechanisms explain the participation decline.

Based on the research findings, this paper proposes three policy recommendations: First, enhance insurance service efficiency. Insurance institutions are advised to systematically improve the enrollment experience by strengthening cross-departmental collaboration, refining safeguard mechanisms, and optimizing product services (Daneshgar et al. 2006). Second, reinforce data security measures. It is essential to strictly comply with privacy regulations and establish comprehensive lifecycle management protocols for sensitive data to ensure the information security of enrollees (Liam & Jun 2007). Third, cultivate demand for old-age security. The government should adopt a dual approach—raising public awareness of retirement planning through policy advocacy while improving institutional education systems to foster market recognition, thereby activating potential demand for commercial pension insurance.

Acknowledgement. This research is supported by the Key Research Platforms and Projects of Ordinary Higher Education Institutions under the Guangdong Provincial Department of Education [NO.SK2024C001], and Shenzhen Institute of Information Technology [NO.2024kcszzx05].

References

Chen, F., Zhai, W.J.: The causes and welfare effects of farmland transfer from the perspective of farmers' behavior. Econ. Res. **50**(10), 163–177 (2015)

Daneshgar, F., Bunker, D., Mawson Lee, K.: Identifying opportunities for strengthening cooperation in the Australian health insurance sector. Int. J. Bus. Process. Integr. Manag. **1**(3), 210–218 (2006)

Kong, X.Z., Xu, Z.Y.: Analysis of factors affecting the choice of transfer objects by farmers transferring out land: An empirical analysis based on a comprehensive perspective. Chin. Rural Econ. (12), 17 - 25 + 67(2010)

Le, Z.: Farmers' willingness to transfer land and its interpretation: An empirical analysis based on the survey data of 1000 farmers in 10 provinces. Issues Agric. Econ. **31**(02), 64 - 70 + 111(2010)

Li, C.X.: Innovative ideas for rural old-age security models under the background of land certification. J. Yunnan Agric. Univ. (Soc. Sci.) **16**(01), 65–70 (2022)

Li, Q., Yang, S.T., Zhang, T.L.: Can social security replace land security? - based on the impact of the new rural pension insurance on the willingness to rent out land rent. Econ. Theory Econ. Manage. **07**, 61–74 (2019)

Li, X.F.: Enlightenment of labor transfer and land transfer models in developed countries to China. Agric. Econ. **06**, 107–109 (2022)

Liam, P., Jun, H.: A service-oriented architecture for managing privacy compliance in collaborative environments. Int. J. Bus. Process. Integr. Manag. **2**(4), 292–301 (2007)

Leng, Z.H., Fu, C.J., Xu, X.P.: Household income structure, income gap and land transfer: a microanalysis based on the data of China Family Panel Studies (CFPS). Econ. Rev. **05**, 111–128 (2015)

Luan, J., Ma, R.: The impact of rural labor transfer employment stability on land transfer: an analysis from the perspective of migration heterogeneity. China Agric. Resources Region. Plan. **42**(12), 203–216 (2021)

Luo, R.F., Liu, Y., Liu, C.F., Zhang, L.X., Zhao, Q.R.: The impact of the new rural pension insurance on farmers' land transfer behavior: Based on the micro-data of 5 provinces from the China Rural Development Survey. Econ. Surv. **36**(03), 33–40 (2019)

Mao, F., Wang, J.J., Kong, X.Z.: The impact of pension insurance participation on farmers' land transfer behavior: an empirical analysis based on CHARLS survey data. Rural Finan. Res. **12**, 21–27 (2020)

Miao, H.M., Zhang, S.L., Zhu, J.F.: The impact of selective migration of migrant workers' families on land transfer: an empirical analysis based on the data of China's floating population dynamic monitoring survey. Chin. Rural Econ. **08**, 24–42 (2021)

Sun, R.T., Xiong, X.P.: The impact of commercial pension insurance on the life satisfaction of rural elderly: survey data from Hubei Province. Develop. Finan. Res. **05**, 45–60 (2019)

Wang, Y.: The perspective of farmers on land transfer and the exploration of transfer paths. Mod. Agric. **01**, 61–63 (2022)

Wu, W.P., Duan, D.Q., Zhou, H.: Economic benefit analysis of pension security on farmland transfer: an empirical analysis based on CHARLS data. Bus. Econ. (09), 132 - 134 + 147 (2020)

Wu, Y.Q., Luo, Q., Mi, C.L., Cai, W.M., Liu, Y.: Gender differences in farmers' willingness to transfer out farmland and influencing factors: an empirical analysis based on 578 questionnaires in Tianjin. China Popul. Resour. Environ. **26**(06), 69–74 (2016)

Xu, J.L., Lu, X.H., Teng, M.L., Zhao, Q.R.: The decision-making logic of rural floating population's land transfer behavior in the western region: from the perspective of push-pull theory. China Agric. Resources Region. Plan. **43**(11), 228–238 (2022)

Xu, Q., Zhou, Y.: The interaction mechanism between family pension and social pension from the perspective of vulnerability: based on the survey data of 1371 households in 8 provinces. J. Jiangxi Univ. Finan. Econ. **05**, 70–80 (2019)

Xu, Z.G., Ning, K., Zhong, F.N., Ji, Y.Q.: The new rural pension insurance and farmland transfer: can institutional pension replace land pension? - based on the perspective of household population structure and liquidity constraints. Manage. World **34**(05), 86 - 97 + 180 (2018)

Yang, H., Peng, K.L.: The impact of household income on land transfer decision-making behavior. J. Jiangsu Agric. Sci. **37**(05), 1320–1326 (2021)

Yu, N., Shi, Q., Jin, H.: Permanent land-use rights and endowment insurance: chinese evidence of the substitution effect. China Econ. Rev. **21**(2), 272–281 (2010)

Zhang, D., Wan, L.: Analysis of factors affecting the transfer of farmers' land contract and management rights: Based on a survey of 15 provinces (regions) in 2004. Chin. Rural Econ. **02**, 24–34 (2007)

Zhang, R.J.: The impact of rural population aging on land transfer: regional differences and comparisons. Agric. Technol. Econ. **9**, 14–23 (2017)

Zhang, Y.F., Lu, Y.: Financial development and land transfer: facts, theory and empirical test. World Agric. **03**, 36–47 (2022)

Zhang, Y.L., Bai, Y.L., Zhen, L., Xin, L.J.: Can the new rural pension insurance promote farmers' land transfer? - based on the three-phase panel data of CHARLS. J. Nat. Resour. **34**(05), 1016–1026 (2019)

Zhang, Y.Q.: The impact of digital inclusive finance on rural land transfer and its mechanism: evidence from CFPS and PKU - DFIIC. Econ. Manage. **36**(03), 30–40 (2022)

Zhang, Y.Q., Chuang, T.H.: Dual challenges of families and land transfer of middle-aged farmers: a mediation test based on the mode and quality of social care services. J. Southwest Minzu Univ. (Hum. Soc. Sci.) **43**(03), 112–123 (2022)

Zhang, Y., Tsai, C.H., Liu, W.: Farmers' policy cognition, psychological constructs and behavior of land transfer: empirical analysis based on household surveys in Beijing. China Agric. Econ. Rev. **15**(2), 323–344 (2023)

Zhu, W., Paudel, K.P., Inoue, S.: Farmland lease, high-rent threat and contract instability: evidence from China. China Agric. Econ. Rev. **13**(4), 799–831 (2021)

Zhu, Y.T., Wu, X.Y.: Research on multi-agent game in the design of rural pension insurance products: based on rural land transfer. China Forestry Econ. **01**, 56–59 (2022)

ClassCube: Effective and Efficient Big OLAP Data Cube Classification via Dimensionality Reduction

Alfredo Cuzzocrea[1,2(✉)], Mojtaba Hajian[1], and Abderraouf Hafsaoui[1]

[1] iDEA Lab, University of Calabria, Rende, Italy
{alfredo.cuzzocrea,mojtaba.hajian,abderraouf.hafsaoui}@unical.it
[2] Department of Computer Science, University of Paris City, Paris, France

Abstract. In this paper, we introduce *ClassCube*, a novel methodology designed to perform *efficient* and *effective* classification over large, multidimensional OLAP data cubes. *ClassCube* leverages *logical cuboid lattices* to represent data across multiple aggregation levels, enabling intelligent selection of both dimensions and cuboids for classification tasks. By integrating dimensionality reduction techniques such as *Dimension Selection* and *Principal Component Analysis*, the approach significantly reduces computational overhead while maintaining high classification accuracy. Extensive experimental assessments confirm that *ClassCube* achieves an optimal balance between *efficiency* and *accuracy*, highlighting its suitability for *real-life big data analytics applications*.

Keywords: Big Data Analytics · Multidimensional Big Data Analytics · Integration of OLAP Analysis and Classification

1 Introduction

In today's data-driven era, organizations across various sectors, such as finance, healthcare, and e-commerce, rely heavily on analyzing large, multidimensional datasets to make strategic decisions. In this context, *Online Analytical Processing* (OLAP) data cubes have become fundamental tools, enabling complex queries and interactive exploration of data aggregate over multiple dimensions. These data cubes facilitate operations like slicing, dicing, and drilling down into data, which are essential for uncovering patterns and trends that lead to business insights.

However, as the volume and dimensionality of data continue to grow exponentially, performing classification tasks directly on OLAP data cubes has become increasingly computationally expensive. High-dimensional data presents significant challenges, especially the *curse of dimensionality*, where the feature space becomes so huge that data

---

This research has been made in the context of the Excellence Chair in Big Data Management and Analytics at University of Paris City, Paris, France

© The Author(s), under exclusive license to Springer Nature Switzerland AG 2025
S. Zhang et al. (Eds.): BIGDATA 2025, LNCS 16152, pp. 179–194, 2025.
https://doi.org/10.1007/978-3-032-06524-7_12

points become sparse. This sparsity adversely affects the performance of classification algorithms, leading to longer computation times and potential overfitting. Consequently, there is a need for efficient techniques that can reduce computational costs while maintaining high performance of classification.

The main challenge focused on in this research regards the substantial computational load associated with performing classification on high-dimensional OLAP data cubes. Traditional classification methods struggle with scalability in such areas due to the extensive resources required for processing and the potential degradation in accuracy caused by the high number of dimensions. This issue is particularly acute in real-time and mobile environments, where computational resources are limited. Therefore, enhancing classification processes to handle high-dimensional data efficiently is crucial for enhancing the effectiveness of data analysis in various applications.

Several research efforts have focused on addressing the computational challenges associated with high-dimensional data classification, particularly in the context of large datasets. *Dimensionality reduction techniques* (e.g., [1]), such as *Principal Component Analysis* (PCA) [2] and *Feature Selection Algorithms* (e.g., [3]), have been widely employed to mitigate the curse of dimensionality. For instance, [4] integrates PCA into a multidimensional F1-transform classifier, effectively reducing computational loads and improving classification accuracy over traditional algorithms. Similarly, [5] develops a *hybrid optimization-based feature selection method* to identify relevant feature subsets, resulting in improved convergence speed and classification performance on high-dimensional datasets. Beyond using dimensionality reduction methods, some studies have explored the integration of these techniques with classification algorithms to further enhance efficiency. [6] introduces *deep self-supervised machine learning* models enriched with novel feature elimination and selection strategies, effectively reducing dimensionality and improving classification accuracy for multidimensional health risk assessments. Additionally, advanced data structuring methods have been proposed to manage high-dimensional data more effectively. [7] presents an *adaptive granularity* and *dimension decoupling network* for multidimensional time series classification, which extracts features at various scales and decouples dimensions to prevent dominant features from overshadowing others.

To tackle the aforementioned challenges, we propose *ClassCube*, a novel methodology that integrates *multidimensional OLAP analysis* [8] and *classification algorithms* [9] to effectively and efficiently support *big data analytics* in real-life scenarios. The key idea of *ClassCube* relies on dimensionality reduction tools (e.g., [1]). At a practical level, we leverage on so-called *big OLAP data cubes* [10] for big data applications, and we address the issue of *effectively and efficiently classifying big multidimensional data* [11] *in Cloud environments* (e.g., [12]). With this goal in mind, we propose the anatomy and main functionalities, including experimental assessment and evaluation, of *ClassCube*, *an innovative methodology for supporting advanced big data analytics via intelligent classification tools over big OLAP data cubes*.

In our study, we focus on leveraging the *logical cuboid lattice*, a hierarchical structure that represents all possible aggregations of the data across different combinations of dimensions. Each cuboid in the lattice corresponds to a specific aggregation level, offering a multi-resolution view of the data. This structure leads the model to more efficient data management and analysis by enabling operations at different levels of granularity. However, even with this hierarchical approach, performing classification directly on the entire lattice remains resource-intensive. It should be noted that the selection process focuses on identifying specific dimensions from the OLAP data cube to define the cuboids of interest. This criterion can be guided by *user/application input* or determined based on *state-of-the-art models* (e.g., [13, 14]).

Our proposed methodology integrates dimensionality reduction with hierarchical data cube structuring to perform an efficient classification process. Specifically, we extract the *logical cuboid lattice* from the original OLAP data cube using *Principal Component Analysis* to construct a new, reduced-dimensionality data cube. PCA effectively reduces the number of dimensions by projecting the most significant features that contribute to data variance, to a reduced feature space. Thus, mitigating the curse of dimensionality. On the other hand, we apply a dimension selection method to extract the reduced-dimensionality data cubes, which then the classification performance is compared with the one resulting from PCA utilization. Thus, the reduced data cube serves as the basis for our classification tasks.

We then perform classification using algorithms such as *Logistic Regression* (LR) and *Support Vector Machines* (SVM) on the cuboids of interest. The reduced dimensionality substantially lowers computational costs while aiming to preserve classification accuracy. Evaluating the effectiveness of this approach presents its own challenges. Direct comparison with the original high-dimensional data cube is impractical due to resource constraints. Therefore, we compare the classification performance on the reduced data cube with that on cuboids from the lattice that have the same dimensionality. This comparative strategy provides a meaningful evaluation for assessing the effectiveness of our methodology.

A preliminary, introductive version of this research appears in [31]. The key contributions of this research are as follows:

- First, we introduce an efficient dimensionality reduction framework that employs a hierarchical data cube structure to be used in classification applications on high-dimensional OLAP data cubes.
- Second, our methodology applies classification algorithms on the cuboids of interest from the extracted *lattice of cuboids*.
- Third, we establish a practical evaluation methodology by comparing the reduced data cubes (i.e., cuboids) from the lattice, providing meaningful insights into the effectiveness of our approach.

2 Related Work

In this Section, we review relevant related work that aligns with our objective of efficient and effective classification of multidimensional OLAP data cubes through the integration of dimensionality reduction techniques.

PCA [2] is one of the most widely used *dimensionality reduction methods*, capable of preserving the most significant variance components. In [4], it is used to enhance the multidimensional F1-transform classifier by reducing data size during preprocessing, which lowers computational costs. Their hybrid approach improves classification accuracy and outperforms traditional classifiers such as *SVM*, *Random Forests* (RF), and *Artificial Neural Networks* (ANN), which demonstrates the effectiveness of PCA in high-dimensional spaces.

Similarly, [5] proposes a *hybrid optimization-based feature selection method* by combining *Slime Mould Algorithm* (SMA) and *Binary Grey Wolf Optimization* (BGWO) to improve classification performance while reducing computational loads.

This approach efficiently identifies relevant features from high-dimensional data, mitigating the curse of dimensionality, and improves *K-Nearest Neighbor* (KNN) classifier results over SVM and Naïve Bayes. [15] employs a *multi-objective genetic algorithm* (GA) guided by feature importance rankings to optimize both accuracy and dimensionality reduction. Their method achieves *Pareto-optimal solutions* and consistently outperforms traditional feature selection techniques across diverse high-dimensional datasets.

Furthermore, the integration of dimensionality reduction techniques with classification algorithms has proven to be an effective strategy in improving performance on high-dimensional datasets. [6] tackles the classification of multidimensional health risks from blood test data using *deep self-supervised learning* enhanced by novel *feature elimination* and *weighting techniques*. They introduce an *adaptive feature elimination method* to reduce redundant data and a *self-feature weighting approach* that prioritizes features deviating from normal values. These enhancements, applied to algorithms like *Batch Least Squares* (BLS) and *Iterative Neural Networks* (INN), improve classification accuracy while reducing computational complexity. On the other hand, [16] addresses the computational challenges of applying *Shapley values* for high-dimensional, time-series data in the context of DNA profile classification using *convolutional neural networks* (CNNs). To overcome this issue, they introduce a novel *partitioning strategy* suitable for time-series data inspired by superpixels in image processing. By segmenting time-series data into meaningful blocks, they enable efficient Shapley value computation without incurring high computational costs.

In the area of *multidimensional functional data*, [17] proposes a *multiclass functional deep neural network* (mfDNN) that combines Functional Data Analysis with deep learning. Using *Functional Principal Component Analysis* (FPCA) for dimensionality reduction, they extract *principal component scores* and input them into a sparse deep neural network with ReLU activation. The mfDNN classifier shows strong classification performance on both simulated and real-life datasets like MNIST and *Alzheimer's Disease Neuroimaging Initiative* (ADNI). [18] presents a deep learning method for classifying epileptic seizures from high-dimensional *Electroencephalography* (EEG) signals. They convert 1D-EEG signals into 2D and 3D spectrograms using *Short-Time Fourier Transform* (STFT) to capture *time-frequency features*. A *deep convolutional autoencoder* (DCAE) then extracts hierarchical features from these spectrograms, enabling effective seizure classification.

In the context of *multidimensional classification* (MDC), [19] introduces a *probability-based label enhancement method* for MDC, where instances belong to multiple heterogeneous class spaces. To address the limitations of traditional methods, they encode class spaces into a *single logical-label space* and minimize the *Kullback-Leibler divergence* between feature and label distribution manifolds. [20] tackles the *Imbalance*

Shift problem in MDC, where class importance varies across labeling dimensions. They propose the *Imbalance-Aware Fusion Model* (IMAM), which decomposes the MDC task into separate *multi-class problems*. Each is handled by an *imbalance-aware deep model* that adjusts prediction logits based on class distributions.

Moreover, *Feature Fusion methods* have been also used in research to mitigate dimensional complexity. [21] addresses the challenge of accurately classifying multiple respiratory diseases by combining *one-dimensional* and *two-dimensional* data representations. By employing widely used 2D representations like *spectrograms* and *mel-frequency cepstral coefficients* (MFCC) alongside raw 1D respiratory data, they create a dataset that includes both *time-series* and *frequency-domain features*. [22] proposes a method that combines multidimensional feature fusion of *Long Short-Term Memory* (LSTM) networks and the *Xception* model for stress classification using *electrocardiograms* (ECGs). They extracted heart rate variability features using LSTM networks and converted ECG signals into spectrogram images to capture spatial features using the Xception model.

[23] addresses the challenges of *Unmanned Aerial Vehicle* (UAV) classification in complex environments by proposing a *deep learning fusion approach* that leverages multiple spectrogram representations derived from radar signals. They integrated *micro-Doppler spectrograms*, *cadence-velocity diagrams*, and *cepstrum images* using two fusion strategies including *data-level fusion* and *feature-level fusion*. Their experimental results demonstrate that both fusion strategies significantly outperform methods that use individual spectrograms or raw data alone. Additionally, [24] addresses the challenges of ranking and classifying multidimensional data by introducing two novel methods including *Spanning Thread* (ST) and *Ordered Spanning Thread* (OST). These methods create *linear orderings* that preserve data point similarities, effectively organizing data without altering dimensionality.

3 *ClassCube* Methodology

In this Section, we present the *ClassCube* methodology for efficient and effective classification of high-dimensional OLAP data cubes by integrating dimensionality reduction techniques with hierarchical data cube structuring. Our approach aims to mitigate computational costs on big OLAP data cube classification while preserving the algorithm's performance.

Consider a big multidimensional OLAP data cube O with N dimensions, denoted as $D = \{d_1, d_2, \ldots, d_N\}$. This data cube contains aggregate data across all possible combinations of these dimensions, simplifying complex analytical queries such as slicing, dicing, and drilling down into data. However, performing classification directly on O is computationally expensive due to the high dimensionality.

To address this challenge, we consider a logical cuboid lattice L from O. The lattice L is a hierarchical structure representing all possible aggregations of O over different combinations of dimensions. Each cuboid within L corresponds to a specific level, providing comprehensive multiple views of the data. By working with cuboids of reduced dimensionality, we aim to alleviate the computational costs associated with high-dimensional data.

To provide more details, a cuboid C is an aggregation of the data over a subset $D' \subseteq D$, where D' contains a specified combination of dimensions. The aggregation

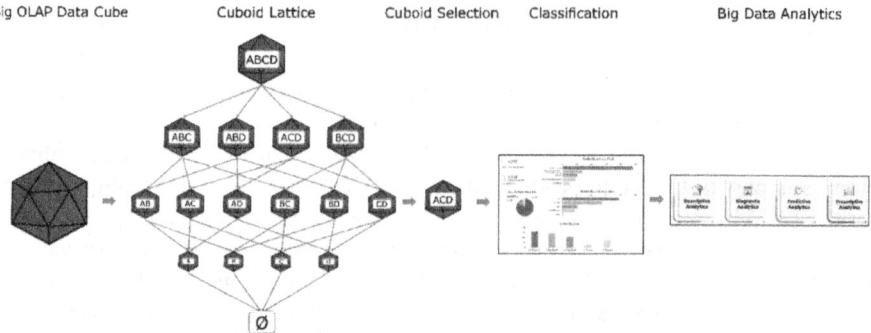

Fig. 1. *ClassCube* Methodology.

within a cuboid may involve operations such as sum, average, or count, over the dimensions in D'. The hierarchical relationship among cuboids is based on the principle of dimension aggregation, where lower-dimensional cuboids are derived by aggregating higher-dimensional ones over one or more dimensions.

Construction of the Lattice Levels: The logical lattice L is defined by hierarchically structured cuboids at different levels as illustrated in Fig. 1. This hierarchical structure enables analysis at various levels of granularity, providing flexibility in the selection of cuboids for classification tasks and addressing challenges related to high dimensionality and computational cost.

To achieve our goal of efficient classification, we propose two approaches to obtain cuboids with the same dimensionality:

- *Dimension Selection*;
- *Principal Component Analysis*.

In both approaches, a specific level k (with dimensionality k_d) of the lattice L is selected using the selection criterion, which is obtained as *user/application input* or determined as previously described based on *state-of-the-art models* (e.g., [13, 14]). Then, a classification algorithm is applied to each cuboid C_{ki} at level k.

Dimension Selection Approach: The dimension selection method involves selecting specific subsets of dimensions from D to create reduced versions of O. For each level k in the lattice L, where $k = 1, 2, \ldots, K$ and $K \leq N$, we define m_k as the number of dimensions at level k (i.e., $m_k = k_d$).

At each level k, we select all possible combinations of m_k dimensions from D. Each combination $D_{ki} \subset D$ defines a cuboid C_{ki}, where $i = 1, 2, \ldots, I_k$, $I_k = \binom{N}{m_k}$ is the number of combinations at level k. The cuboid C_{ki} is obtained by aggregating O over the dimensions included in D_{ki}.

Therefore, the set of cuboids at level k is as follows:

$$L_k = \{C_{ki} | D_{ki} \subset D, |D_{ki}| = m_k\} \qquad (1)$$

Each cuboid C_{ki} has dimensionality m_k and provides a view of the data over m_k (=k_d) dimensions. By working with this lower-dimensional cuboid, we reduce the computational complexity of the classification task.

Cuboid Representation Using Principal Component Analysis (PCA). In our second approach, we use PCA to obtain a reduced-dimensionality representation of the data cube O with the same dimensionality k_d as in the dimension selection approach. PCA is applied to O to reduce its dimensionality by transforming the original data into a new set of orthogonal components that capture the maximum variance.

We compute the covariance matrix $\sum$ of O and solve for its eigenvalues and eigenvectors. The top k_d eigenvectors corresponding to the largest eigenvalues form the transformation matrix P. The reduced data cube C_{PCA} is obtained by projecting O onto the new feature space which provides C_{PCA} as follows:

$$C_{PCA} = O \times P \tag{2}$$

This projection results in a single reduced-dimensionality representation, capturing the most significant patterns in the data. By comparing the classification performance using the cuboids from the dimension selection approach and the PCA-reduced data cube, we evaluate the effectiveness of our methodology.

Classification Algorithms: we apply classification algorithms to the selected cuboids S to evaluate the effectiveness of our dimensionality reduction approach. Specifically, we use *Logistic Regression* and *Support Vector Machines*, which are well-suited for handling reduced-dimensionality data.

Logistic Regression models the probability $P(y = 1|x)$ of a binary outcome using the logistic function as follows:

$$P(y = 1|x) = \frac{1}{1 + e^{-(\beta_0 + \beta^T x)}} \tag{3}$$

where β_0 is the intercept, β is the coefficient vector, and x is the feature vector from a cuboid C_{ki}. The parameters β_0 and β are estimated using *Maximum Likelihood Estimation* (MLE). The likelihood function for a set of observations $\{(x_i, y_i)\}, i = 1, 2, \ldots, n$ is given by:

$$L(\beta_0, \beta) = \prod_{i=1}^{n} P(y_i|x_i)^{y_i}[1 - P(y_i|x_i)]^{1-y_i} \tag{4}$$

Maximizing the likelihood function (or equivalently, the log-likelihood) obtains estimates of the parameters that best fit the observed data. The optimization is typically performed using iterative algorithms like Newton-Raphson or gradient descent. To further prevent overfitting, regularization methods can be imposed into the LR model. The common regularization terms are as follows:

- *L1 Regularization (Lasso Regression)*: Adds a penalty equal to the absolute value of the magnitude of coefficients. It performs feature selection by shrinking some coefficients to zero.
- *L2 Regularization (Ridge Regression)*: Adds a penalty equal to the square of the magnitude of coefficients. It prevents large coefficients but does not enforce sparsity.

The regularized cost function becomes as follows:

$$J(\beta_0, \beta) = -\frac{1}{n}\sum_{i=1}^{n}[y_i \log P(y_i|x_i) + (1-y_i)\log(1-P(y_i|x_i))] + \lambda R(\beta) \quad (5)$$

where $R(\beta)$ is the regularization term and λ is the regularization parameter controlling the trade-off between fitting the data and keeping the model coefficients small.

Support Vector Machines are powerful supervised learning models used for classification and regression tasks. They are particularly effective in high-dimensional spaces and are robust against overfitting, especially in cases where the number of dimensions exceeds the number of observations. SVMs are well-suited for our methodology, which involves reduced-dimensionality data cubes derived from high-dimensional OLAP systems.

SVM algorithm seeks an optimal hyperplane that maximally separates data points of different classes. For linearly separable data, the objective is to maximize the margin between the closest points (support vectors) of the two classes. The optimization problem is formulated as:

$$\min_{w,b} \frac{1}{2}\|w\|^2 \quad (6)$$

$$s.t. \ y_i\left(w^T x_i + b\right) \geq 1, \forall i$$

where w is the normal vector to the hyperplane, b is the bias term, x_i are the feature vectors, and $y_i \in \{-1,1\}$ are the class labels. By solving this convex optimization problem, SVM identifies the hyperplane that not only separates the classes but does so with the greatest possible margin, enhancing the model's generalization capabilities.

Real-world data often cannot be perfectly separated by a linear hyperplane. To address this, SVM introduces the concept of kernel functions. Kernel function projects the original feature space into a higher-dimensional space where linear separation is possible. Common kernels include:

- *Linear Kernel*: $K(x_i, x_j) = x_i^T x_j$.
- *Polynomial Kernel*: $K(x_i, x_j) = (\gamma x_i^T x_j + r)^d$.
- *Radial Basis Function* (RBF) *Kernel*: $K(x_i, x_j) = exp(-\gamma \|x_i - x_j\|^2)$

Selecting appropriate hyperparameters is crucial for SVM performance: (*i*) *Regularization Parameter C*: Determines the penalty for misclassification. A large C prioritizes classification accuracy on the training data, potentially at the expense of generalization. (*ii*) *Kernel Parameters*: Parameters like γ in the RBF kernel or d in the polynomial kernel affect the flexibility of the decision boundary. (*iii*) *Cross-Validation*: Techniques such as k-fold cross-validation are used to systematically explore combinations of hyperparameters to identify the optimal model settings.

4 Case Study: Virus Classifications in Healthcare Data

In modern healthcare, accurate and on-time identification of viral infections is essential for patient management, epidemiological tracking, and containment efforts. However, healthcare data often consists of large, high-dimensional datasets with multiple features like patient demographics, clinical symptoms, and various biomarkers. Applying traditional classification techniques directly on such large datasets can be computationally expensive and inefficient, especially when rapid responses are required.

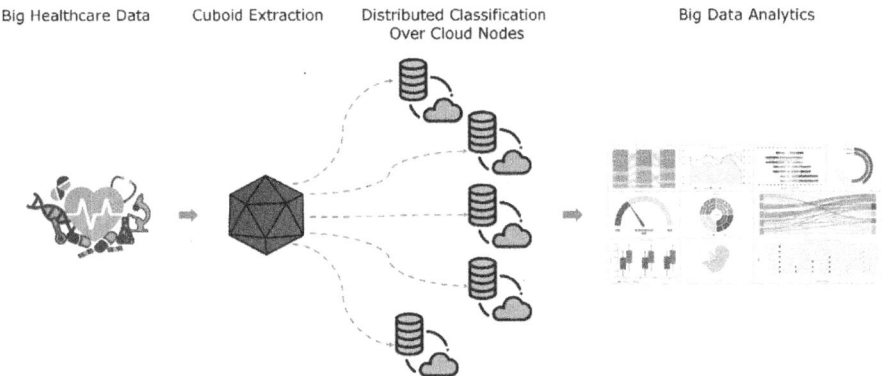

Fig. 2. Case Study: Virus Classification in Healthcare Data.

In this case study, we demonstrate the application of our classification method on a healthcare dataset focused on virus classification, illustrating the method's effectiveness and efficiency through Cloud-based big data analytics. An overview of the architecture of our proposed approach is shown in Fig. 2. In the following, we provide more details about the approach using a healthcare case study.

Healthcare Data Structure: the first stage of our approach involves preparing the high-dimensional healthcare data for classification. In our virus classification example, the dataset consists of numerous dimensions, including patient information (e.g., age, medical history), viral load, and various biomarkers (e.g., antibody levels, protein expressions) that provide distinguishing characteristics among different virus types, such as influenza or coronavirus. The data is structured in an OLAP format, enabling complex multidimensional analysis essential for healthcare applications.

Cuboid Extraction: in this stage, we decompose the high-dimensional OLAP data cube O into multiple lower-dimensional cuboids C_{ki}, each representing an aggregate of the data across specific subsets of dimensions. This process, referred to as cuboid extraction (as illustrated in Fig. 2), uses the hierarchical cuboid lattice L to provide analysis at various levels of granularity.

Formally, let $D = \{d_1, d_2, \ldots, d_N\}$ denote the set of all dimensions in the original data cube O. Each cuboid C_{ki}, is defined by a subset of dimensions $D_{ki} \subset D$, where $|D_{ki}| = N - k$ and $k \in \{0, 1, 2, \ldots, N\}$ referred to the level number. The total number of possible cuboids at level k is given by the combination $\binom{N}{N-k}$ representing all combinations of k dimensions from D. Each cuboid C_{ki} is obtained by aggregating the original data cube O over the dimensions not included in D_{ki}. The aggregation operation can be formalized as follows:

$$C_{ki} = AGG(O, \{d \in D | d \notin D_{ki}\}) \tag{7}$$

where AGG represents an aggregation function such as SUM, AVG, or $COUNT$ over the specified dimensions. The set of cuboids at level k is then defined as L_k. By decomposing O into cuboids C_{ki}, we generate a set of smaller, more manageable data cubes that capture specific views of the data. For example, in our healthcare case study, one cuboid might aggregate patient data over demographic dimensions (e.g., *age*, *gender*), while another focuses on specific biomarkers (e.g., *antibody levels*, *protein expressions*).

To evaluate our approach with another dimensionality reduction method, we apply PCA to reduce the dimensionality of the original OLAP data cube O. PCA projects the data onto a lower-dimensional space by retaining the principal components that capture the most significant variance. This process helps in simplifying complex data structures, ensuring that the most informative features are obtained in each cuboid. Therefore, through the cuboid extraction process, whether via dimension selection or PCA, we obtain a set of cuboids $\{C_{ki}\}$ or a reduced data cube C_{PCA} that are suitable for efficient classification and analysis in subsequent stages. By reducing dimensionality, we alleviate the computational load required for classification, making it possible to handle each cuboid independently in a distributed environment.

Distributed Classification over Cloud Nodes: in the third stage, we select a specific cuboid C_{k*} from the set of cuboids $\{C_{ki}\}$ based on a predefined criterion, namely Φ, and distribute this cuboid over the Cloud nodes for parallelized classification tasks. Consider $C = \{C_{ki} | D_{ki} \subset D, |D_{ki}| = N - k\}$, then this procedure is defined as follows:

$$C_{k*} = \Phi(C) \tag{8}$$

Once the specific cuboid C_{k*} is selected, it can be distributed on the Cloud environment for classification tasks, as shown in Fig. 2. Thus, C_{k*} is partitioned into M subsets to be distributed over the Cloud nodes $v = \{n_1, n_2, \ldots, n_M\}$. The partitioning is done as follows:

$$\bigcup_{m=1}^{M} C_m = C_{ki}, (C_m \cap C_{m'} = \emptyset, \forall m \neq m) \tag{9}$$

Each Cloud node n_m processes its assigned subset C_m independently, applying classification algorithms such as LR and SVM. For virus classification, LR calculates the probability of each virus type based on the cuboid's features, while SVM identifies a hyperplane that maximally separates the virus classes within the reduced feature space. By operating on reduced-dimensional cuboids, both algorithms are able to perform efficiently without encountering the computational challenges typically associated with high-dimensional data especially when enriched by parallel processing over the Cloud nodes. This approach enables our model to scale with large healthcare datasets, maintaining low latency and enabling real-time analysis.

Big Data Analytics: the final step in our approach, Big Data Analytics, involves aggregating and analyzing the classification results from all Cloud nodes. By combining the output from each node, we obtain comprehensive analytics on virus classification across the entire healthcare dataset. Our approach also leverages multi-resolution analysis enabled by the cuboid lattice, which means that virus classification can be evaluated at different aggregation levels. This flexibility gives us the ability to capture insights at varying levels of detail.

5 Experimental Results

In this Section, we present the results of our experiments conducted on a 6D OLAP data cube built on top of the state-of-art dataset *Adult* [25]. Particularly, we employed two dimensionality reduction techniques, namely Dimension Selection approach, and PCA. We evaluated the classification performance using LR and SVM on the reduced data cubes and compared these results together as well as the performance on the entire OLAP cube. Key performance metrics include *Accuracy, Confusion Matrices,* and *ROC curves with AUC values.*

Our first aim was to evaluate the efficacy of reducing the dimensionality of the OLAP data cube from 6D to 4D. We applied two key dimensionality reduction techniques: PCA and Dimension Selection, and employed LR and SVM to classify the data.

Starting with LR, we observe in Fig. 3(a) that the performance on the original 6D data cube is well with AUC values nearing 1.0 for most classes, peaking at 0.99732. The corresponding confusion matrix (see Fig. 3(b)) illustrated a reasonable classification accuracy, demonstrating the model's effectiveness when all six dimensions were utilized.

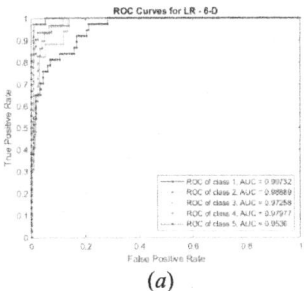
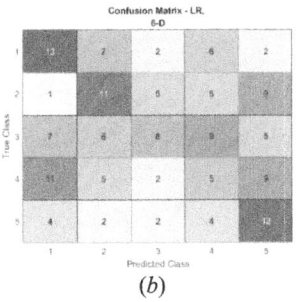

(a) (b)

Fig. 3. AUC Curves (*a*) – Confusion Matrix (*b*) for Logistic Regression over 6D Data Cube.

Similarly, the SVM classifier shows good performance on the 6D data cube (see Fig. 4), with AUC values slightly lower than those of Logistic Regression, reaching up to 0.97977.

However, when we reduced the data cube to 4D using PCA, the LR classifier continued to deliver good results. As shown in Fig. 5a, the AUC values ranged from 0.93835 to 0.99858, with only a slight decline in performance for some classes. The confusion matrix (see Fig. 5b) reveals a marginal increase in misclassifications, but overall, the accuracy remained strong. This indicates that PCA successfully captured the most significant features of the data, enabling the 4D model to perform almost as well as the 6D model. The small decrease in classification power is a reasonable trade-off considering the reduced computational complexity, suggesting that the 4D model is a highly efficient alternative.

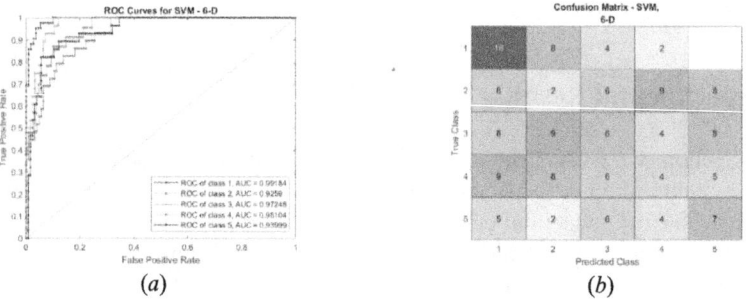

Fig. 4. AUC Curves (*a*) – Confusion Matrix (*b*) for SVM over 6D Data Cube.

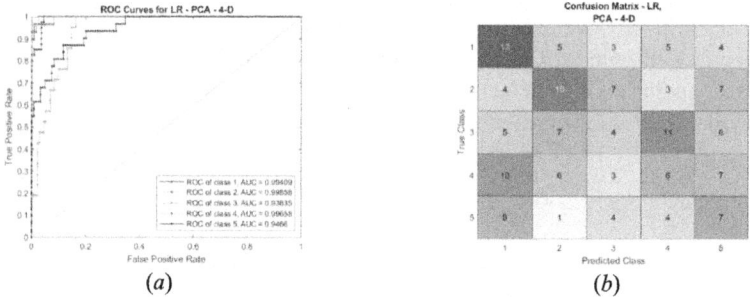

Fig. 5. AUC Curves (*a*) – Confusion Matrix (*b*) for Logistic Regression with PCA over 4D Data Cube.

Additionally, the confusion matrix shows that SVM effectively classified most data points, though it struggled with certain overlapping classes. When the dimensionality was reduced to 4D using PCA, the AUC values for SVM ranged from 0.90146 to 0.98541 (see Fig. 6a). Although there was a more noticeable drop in performance for some classes compared to Logistic Regression, the overall classification accuracy remained satisfactory. The confusion matrix (see Fig. 6b) highlighted more significant challenges in class separation, but given the substantial decrease in complexity, the 4D SVM model still offers an efficient solution.

The Dimension Selection approach followed a similar trend. Using Logistic Regression, when the dimensionality was reduced to 4D using Dimension Selection, the performance was comparable to that of PCA. As shown in Fig. 7a, the AUC values were slightly lower, but the overall accuracy remained robust. The confusion matrix (see

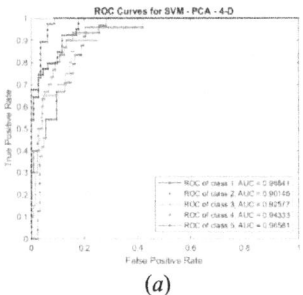 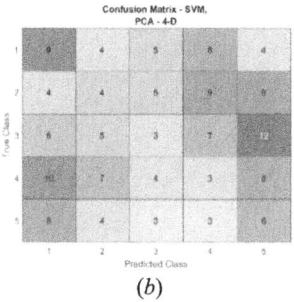

Fig. 6. AUC Curves (*a*) – Confusion Matrix (*b*) for SVM with PCA over 4D Data Cube.

Fig. 7b) indicated a few additional misclassifications, but the results were still reasonable, demonstrating that the selected dimensions preserved critical information for classification. This result reinforces the value of the Dimension Selection method, particularly in scenarios where interpretability is crucial, as it provides an understandable set of features without sacrificing much performance.

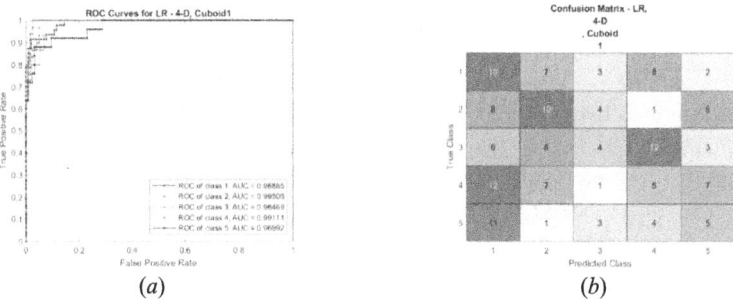

Fig. 7. AUC Curves (*a*) – Confusion Matrix (*b*) for Logistic Regression with Dimension Selection over 4D Data Cube.

Regarding the SVM classifier, when combined with Dimension Selection, there was a slight decrease in performance when dimensionality was reduced to 4D. As shown in Fig. 8a, the AUC values ranged from 0.90346 to 0.96861, and the confusion matrix (see Fig. 8b) illustrates increased misclassifications for certain classes.

The comparison between the 4D cuboid and the 6D original OLAP cube shows that dimensionality reduction has minimal impact on classification performance, especially for Logistic Regression. The slight drop in accuracy is acceptable considering the reduced complexity. This demonstrates that the approach effectively mitigates the curse of dimensionality while preserving predictive power and enabling efficient Cloud-based deployment.

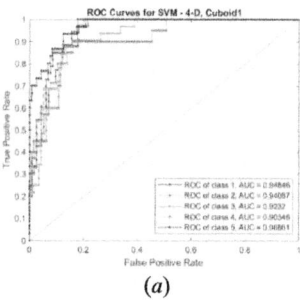 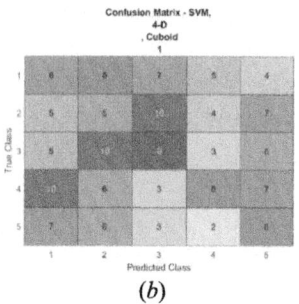

Fig. 8. AUC Curves (*a*) – Confusion Matrix (*b*) for SVM with Dimension Selection over 4D Data Cube.

6 Conclusions and Future Work

In this paper, we propose and experimentally assess *ClassCube*, an innovative methodology for effective and efficient big OLAP data cube classification via dimensionality reduction techniques. Our proposed approach leverages *logical cuboid lattices* to represent data at multiple aggregation levels, which enables efficient selection of dimensions and cuboids for classification tasks. The experimental results highlight the trade-off between *reduced computational overhead* and maintained *classification accuracy*. The use of *Logistic Regression* and *SVM* on reduced-dimensional cuboids highlights that our approach effectively preserves performance while significantly reducing resource requirements.

Future work is mainly oriented through extending our methodology with advanced *Machine Learning* models to further enhance *flexibility* and *scalability* as well as integrate emerging *big data trends* (e.g., [26–30]).

Acknowledgments. This work was funded by the Next Generation EU - Italian NRRP, Mission 4, Component 2, Investment 1.5 (Directorial Decree n. 2021/3277) - project Tech4You n. ECS0000009.

References

1. Sorzano, C.O.S., Vargas, J., Pascual-Montano, A.P.: A survey of dimensionality reduction techniques. CoRR abs/1403.2877 (2014)
2. Abdi, H., Williams, L.J.: Principal component analysis. WIREs Comput. Stat. **2**(4), 433–459 (2010)
3. Molina, L.C., Belanche, L., Nebot, A.: Feature selection algorithms: a survey and experimental evaluation. In: 2002 IEEE ICDM, pp. 306–313 (2002)
4. Cardone, B., Di Martino, F.: A novel classification algorithm based on multidimensional F1 fuzzy transform and PCA feature extraction. Algorithms **16**(3), 128 (2023)
5. Khan, A.A.R., Nisha, S.S.: Efficient hybrid optimization-based feature selection and classification on high dimensional dataset. Multim. Tools Appl. **83**(20), 58689–58727 (2024)

6. Tutsoy, O., Koç, G.G.: Deep self-supervised machine learning algorithms with novel feature elimination and selection approaches for blood test-based multi-dimensional health risks classification. BMC Bioinform. **25**(1), 103 (2024)
7. Ding, G., Geng, S., Jiao, Q., Jiang, T.: AGDM: Adaptive Granularity and Dimension Decoupling for Multidimensional Time Series Classification. In: *13th ICIC*, pp. 405–416 (2024)
8. Gray, J., et al.: Data Cube: A Relational Aggregation Operator Generalizing Group-by, Cross-Tab, and Sub Totals. Data Min. Knowl. Disc. **1**(1), 29–53 (1997)
9. Hassan, C.A.U., Khan, M.S., Shah, M.A.: Comparison of machine learning algorithms in data classification. In: 24th IEEE ICAC, pp. 1–6 (2018)
10. Cuzzocrea, A.: Big OLAP data cube compression algorithms in column-oriented cloud/edge data infrastructures. In: 9th IEEE BigMM, pp. 1–2 (2023)
11. Cuzzocrea, A., Song, I.Y., Davis, K.C.: Analytics over Large-Scale Multidimensional Data: The Big Data Revolution! In: 14th ACM DOLAP, pp. 101–104 (2011)
12. Nodarakis, N., Sioutas, S., Tsoumakos, D., Tzimas, G., Pitoura, E.: Rapid AkNN query processing for fast classification of multidimensional data in the cloud. CoRR abs/1402.7063 (2014)
13. Lin, W.Y., Kuo, I.C.: A genetic selection algorithm for OLAP data cubes. Knowl. Inf. Syst. **6**(1), 83–102 (2004)
14. Talebi, Z.A., Chirkova, R., Fathi, Y., Stallmann, M.F.: Exact and inexact methods for selecting views and indexes for OLAP performance improvement. In: 11th ACM EDBT, pp. 311–322 (2008)
15. Bohrer, J.S., Dorn, M.: Enhancing classification with hybrid feature selection: a multi-objective genetic algorithm for high-dimensional data. Expert Syst. Appl. **255**, 124518 (2024)
16. Elborough, L., Taylor, D., Humphries, M.: A novel application of shapley values for large multidimensional time-series data: applying explainable AI to a DNA profile classification neural network. CoRR abs/2409.18156 (2024)
17. Wang, S., Cao, G.: Multiclass classification for multidimensional functional data through deep neural networks. CoRR abs/2305.13349 (2023)
18. Mutersbaugh, J., Lam, V., Linguraur, M.G., Anwar, S.M.: Epileptic seizure classification using multidimensional EEG spectrograms. In: 19th IEEE SIPAIM, pp. 1–4 (2023)
19. Tang, J., Chen, W., Wang, K., Zhang, Y., Liang, D.: Probability-based label enhancement for multi-dimensional classification. Inf. Sci. **653**, 119790 (2024)
20. Shi, Y., Ye, H.J., Man, D., Han, X., Zhan, D.C., Jiang, Y.: Revisiting multi-dimensional classification from a dimension-wise perspective. Front. Comput. Sci. **19**(1), 191304 (2025)
21. Kim, Y., Camacho, D., Choi, C.: Real-time multi-class classification of respiratory diseases through dimensional data combinations. Cognit. Comput. **16**(2), 776–787 (2024)
22. Song, C.H., Kim, J.S., Kim, J.M., Pan, S.B.: Stress classification using ECGs based on a multi-dimensional feature fusion of LSTM and xception. IEEE Access **12**, 19077–19086 (2024)
23. Chen, X., Ma, C., Zhao, C., Luo, Y.: UAV classification based on deep learning fusion of multidimensional UAV micro-doppler image features. IEEE Geosci. Remote Sens. Lett. **21**, 1–5 (2024)
24. Hussenet, L., Boucetta, C., Herbin, M.: Spanning thread: a multidimensional classification method for efficient data center management. In: 24th I4CS, pp. 219–234 (2024)
25. Becker, B., Kohavi, R.: Adult dataset. UCI Machine Learning Repository (1996)
26. Cuzzocrea, A., Furfaro, F., Mazzeo, G.M., Saccà D.: A grid framework for approximate aggregate query answering on summarized sensor network readings. In: 2004 OTMW, pp. 144–153 (2004)

27. Cuzzocrea, A.: Improving range-SUM query evaluation on data cubes via polynomial approximation. Data Knowl. Eng. **56**(2), 85–121 (2006)
28. Cuzzocrea, A., Saccà, D., Serafino, P.: Semantics-aware advanced OLAP visualization of multidimensional data cubes. Int. J. Data Warehous. Min. **3**(4), 1–30 (2007)
29. Cuzzocrea, A.: CAMS: OLAPing multidimensional data streams efficiently. In: 11th DaWaK, pp. 48–62 (2009)
30. Yu, B., Cuzzocrea, A., Jeong, D.H., Maydebura, S.: On managing very large sensor-network data using bigtable. In: 12th IEEE/ACM CCCGrid, pp. 918–922 (2012)
31. Cuzzocrea, A., Hajian, M.: Towards big OLAP data cube classification methodologies: the ClassCube framework. In: 27th ICEIS, pp. 351–356 (2025)

Estimation of Channel Parameters Based on Multilayer Perceptron and Residual Blocks over Rician Fading Channels

Wen-Long Chin[(✉)], Li-Cheng Lo, Yu-Xiang Huang, and Cheng-Hsien Yu

Department of Engineering Science, National Cheng Kung University, Tainan, Taiwan
wlchin@ncku.edu.tw, {n96131590,n96131037}@gs.ncku.edu.tw

Abstract. In this work, three key parameters of the channel, Doppler frequency, Rician factor, and angle-of-arrival (AoA), are estimated by analyzing the correlation characteristics of the received signals and combining them with the residual block technique through the Multilayer perceptron (MLP) model. The final estimation of the channel parameters is validated through simulations and the performance is good under various environmental variations.

Keywords: Angle-of-arrival · Doppler frequency · multilayer perceptron · residual block · rician factor

1 Introduction

Signals traversing different paths can cause interference and delay, resulting in multipath effects. These effects may lead to varying degrees of signal distortion or attenuation at the receiver. Therefore, a thorough understanding of multipath propagation and the associated channel parameter variations is essential for effectively enhancing and restoring communication quality.

Several methods have been proposed for estimating Doppler frequency in mobile communication systems. In [4], a Doppler frequency bias estimation method for Rayleigh fading channels is presented. It derives closed-form weighted least squares equations using basis expansion approximations. The results demonstrate that this method achieves high accuracy in Doppler frequency estimation. A two-step Doppler frequency estimation approach is proposed in [8]. The core idea is to estimate the Doppler frequency deviation using received signals from an Orthogonal Frequency-Division Multiplexing (OFDM) system, aiming to mitigate intercarrier interference (ICI). While this method is effective in reducing ICI and accurately estimating frequency deviations, it is computationally expensive due to the large search space required to find the optimal value.

In [14], the Expectation-Maximization (EM) algorithm and the Alternating ProjectionâĂŞMaximum Likelihood Estimation (AP-MLE) method are employed to jointly estimate Doppler frequency. The AP-MLE algorithm provides improved approximation of the CramérâĂŞRao Lower Bound (CRLB) compared to traditional methods. However, this comes at the cost of high computational complexity. To address this, the integration of EM and AP-MLE is proposed—where EM reduces the number of iterations, albeit still incurring significant computational overhead. Lastly, [1] is the first to investigate non-data-aided Doppler frequency estimation for OFDM systems over doubly selective line-of-sight (LOS) and non-line-of-sight (NLOS) channels. The study proposes a practical simplified estimator and provides an analysis of the corresponding CRLB.

To estimate the angle of arrival (AoA), [6] utilizes the first arrival path (FAP), which is identified through coarse frame synchronization and frequency-domain channel estimation. Once the FAP is located, the AoA can be determined accordingly. In [10], a subspace-based algorithm is proposed to separate the signal and noise subspaces for efficient AoA estimation. This method is conceptually similar to the well-known Multiple Signal Classification (MUSIC) algorithm, which distinguishes signal-space eigenvalues from noise-space eigenvalues. The approach employs singular value decomposition (SVD) to enhance separation accuracy. SVD addresses key limitations of the traditional MUSIC algorithm—particularly its restriction on the number of resolvable AoA components due to the limited number of antennas. Additionally, the use of SVD improves the algorithm's ability to handle channels with multiple AoA components.

The Rician factor in [11] is estimated by comparing the changes in the probability density function (PDF) of the received signals with an established database. Study [12] derived a closed-form expression to estimate the Rician factor. This is achieved by firstly deriving the mean and the variance of the received signals, and then further deriving the closed-form expression and simulating it by the Monte Carlo method. The accuracy of the Rician factor is superior to that of the conventional estimator in terms of signal-to-noise ratio (SNR) and the various variations of AoA. Another crucial aspect of wireless communication is the estimation of channel state information (CSI).

A convolutional neural network (CNN) combined with a long short-term memory (LSTM) has been proposed in [9] as a solution to this problem. Reference [7] presented a novel joint classification method based on a CNN, utilizing spectrogram images to characterize the spurious power ratio and Doppler frequency deviation. This approach is designed to achieve efficient adaptive modulation and coding (AMC). Gated-convolutional neural networks (GCNN), investigated in [2], are more suitable for indoor localization and achieve smaller root-mean-square error (RMSE) compared to conventional CNNs. A pipelined machine learning (ML) algorithm is proposed in [3] to decompose the original ML model into several smaller models, effectively reducing the computational burden through simulation verification.

This work proposes a method that integrates the Multilayer perceptron (MLP) model with residual blocks for application in complex communication systems. Inspired by the concept of biological neurons, this approach aims to estimate three crucial parameters in Rician fading channels. By estimating the Doppler frequency offset, AoA, and Rician factor, the current communication quality can be applied to assess the relative motion speed and position of objects. This holds significant relevance for numerous application scenarios, particularly in harsh environments or situations requiring high precision positioning.

The remainder of this article can be summarized as follows. Section 2 explores MLP and residual blocks. Section 3 introduces the signal model and signal processing process, and then constructs the data set through time diversity and expansion approximation. Section 4 explains the model architecture used, the method of combining residual blocks, and the input data used. Section 5 simulates four different changes in the channel caused by environmental influences and compares the adaptability and performance of the method proposed in this work with that of the traditional method. Section 6 summarizes the advantages and related applications of the method proposed in this work.

2 Multilayer Perceptron

A multilayer perceptron (MLP) is a type of feedforward neural network composed of an input layer, one or more hidden layers, and an output layer. Each neuron in a given layer is fully connected to all neurons in the preceding layer. The network employs weights, biases, and nonlinear activation functions to transform input data. During training, MLPs use backpropagation and gradient descent algorithms to minimize the error between predicted and actual outputs. This enables the network to capture complex nonlinear relationships. MLPs are particularly effective for tasks that require modeling intricate nonlinear mappings, such as classification and regression.

A residual block is a fundamental building block in deep neural networks, based on the concept of residual learning introduced by Kaiming He et al. in 2016 [5]. In traditional deep architectures, increasing network depth often leads to degraded performance, where training error plateaus or even increases. This issue is primarily caused by the vanishing gradient problem. To address this, residual blocks introduce a shortcut (or residual) connection, as illustrated in Fig. 1, which allows the network to learn residual functions instead of full mappings. Specifically, the block outputs the sum of its input and the learned residual, effectively propagating the difference between the input and the desired output. This design simplifies optimization and enables the training of much deeper networks, leading to improved model performance. As a result, residual blocks have become one of the most influential innovations in deep learning, widely adopted in visual recognition and many other applications.

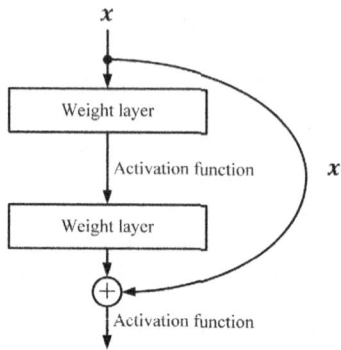

Fig. 1. A Residual Block Connection skips two layers.

3 Signal Model and Data Collection

Wireless communication channels can be categorized into two types: LOS and NLOS, as shown in Fig. 2. LOS channels have an unobstructed path between the transmitter and receiver, while NLOS channels are obstructed by obstacles. LOS channels mainly consist of the direct path between transmitter and receiver, while NLOS channels involve multiple scattering paths. Signals traversing disparate paths will result in interference and delay. This phenomenon can lead to varying degrees of signal distortion or attenuation at the receiver.

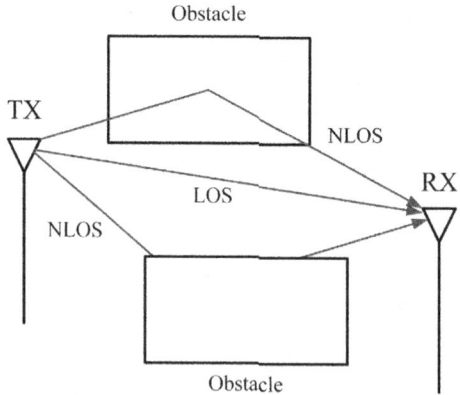

Fig. 2. Various propagation paths.

The channel environment utilized in this work is the Rician fading channel, in which the signal is attenuated and delayed to varying degrees due to scattering, an effect known as the multipath propagation effect or delay spread. The delay when the signal reaches the receiver through different paths is represented by $l_0, l_1, \ldots, l_{L-1}, l_L$. The channel impulse response (CIR) is described as

$$h(t,\tau) = \sum_{l=0}^{L} h_l(t) \cdot \delta(\tau - \tau_l) \qquad (1)$$

In Eq. (1), it is assumed that there are $L+1$ resolvable multipath signals, τ_l denotes the excess delay, with $\tau_l = lT_s$, where T_s is the sampling period. $\delta(\cdot)$ represents the Dirac delta function.

The direct wave component $h_0(t)$ can be expressed as

$$h_0(t) = \sigma_{h(0)} \left(\sqrt{K_0} \rho_0(t) + \sqrt{1 - K_0} \mu_0(t) \right) \qquad (2)$$

In Eq. (2), $\sigma_{h(0)}$ represents the standard deviation of the first path channel tap; $\rho_0(t)$ denotes the LOS component; $\mu_0(t)$ represents the NLOS component; $K_0 \equiv (\frac{\kappa_0}{1+\kappa_0})$ is used to express the power ratio between the LOS and NLOS components, where κ_0 is the Rician factor. If $\kappa_0 = 0$, then $K_0 = 0$, and in this case, Eq. (2) reduces to the NLOS component, implying degradation to a Rayleigh fading channel. The LOS component is related to the angle and velocity of the transmitting and receiving devices, thus the LOS component can be expressed as $\rho_0(t) = e^{j(2\pi f_d \cos(\theta_0)t + \phi_0)}$, where f_d is the maximum Doppler frequency $f_d \equiv v \cdot (\frac{f_c}{c})$, with v being the velocity of the device in meters per second; f_c is the carrier frequency; c is the speed of light; θ_0 is the AoA; ϕ_0 is the random phase angle from $-\pi \sim \pi$.

The NLOS channel tap $h_l(t)$ can be described as

$$h_l(t) = \sigma_{h(l)} \mu_l(t), l \neq 0 \qquad (3)$$

In Eq. (3), $\sigma_{h(l)}$ represents the standard deviation of $h_l(t)$, while $\mu_l(t)$ denotes the Gaussian process. If each channel tap is a wide-sense stationary (WSS) process, the cross correlation of channel taps can be expressed as

$$\begin{aligned} R_{h_l}(t_\Delta) &\equiv E[h_l(t_1) h_l^*(t_2)] \\ &= \sigma_{h(l)}^2 K_l \left(e^{j(2\pi f_d \cos(\theta_0) t_\Delta)} - J_0(2\pi f_d t_\Delta) \right) \\ &\quad + \sigma_{h(l)}^2 J_0(2\pi f_d t_\Delta) \end{aligned} \qquad (4)$$

In Eq. (4), $E[\cdot]$ represents the statistical expectation, t_Δ denotes the time difference, $t_\Delta = t_2 - t_1$, $(\cdot)^*$ represents the conjugate operator, $J_0(\cdot)$ denotes the Bessel function of the first kind of order zero [13].

The transmission signal is generated using a complex Gaussian random program $x[n]$

$$x[n] = \rho x[n-1] + \sqrt{1-\rho^2} z[n] \qquad (5)$$

In Eq. (5), ρ represents the correlation coefficient between $x[n]$ and $x[n-1]$, with $0 < \rho < 1$, $z[n]$ is a Gaussian distribution.

After the transmitted signal went through the multipath channel, it could be influenced by additive white Gaussian noise (AWGN), so the received signal can be indicated as

$$y[n] = x[n] \otimes h[n,l] + z[n] \tag{6}$$

$$h[n,l] \equiv h(t,\tau)|_{t=nT_s, \tau=lT_s} \tag{7}$$

In Eq. (6), $\otimes$ represents the convolution operator, $h[n,l]$ denotes the channel impulse response. Since $x[n]$, $h[n,l]$, and $z[n]$ are uncorrelated with each other. Furthermore, the autocorrelation of $y[n]$ can be written as

$$\gamma_y[n,m] \equiv y[n]y^*[n+m] \tag{8}$$

$$E[\gamma_y[n,m]] = R_x[m]\sum_{l=0}^{L} R_{h_l}[m] + \sigma_z^2 \delta[m] \tag{9}$$

In Eq. (9), m represents the time interval, $R_x[m] \equiv E[x[n]x^*[n+m]]$. Since $z[n]$ is an additive Gaussian white noise. Therefore, it can be simplified to $E[z[n]z^*[n+m]] = \sigma_z^2 \delta[m]$.

Prior to constructing the dataset, the concept of time diversity is introduced to define the relationship between the received samples of the viewing window.

$$\Gamma_y[m] \equiv \frac{1}{N}\sum_{n=0}^{N-1}\gamma_y[n,m] \tag{10}$$

In Eq. (10), N represents the number of samples associated with the sample. According to the Central Limit Theorem (CLT), when $N \to \infty$, $\Gamma_y(m)$ will gradually approach $E[\gamma_y[n,m]]$, and it should be noted that the observation window in the channel is fixed when the time interval is not 0. According to Eq. (9), this can be written as

$$E[\Gamma_y[m]] \approx \frac{1}{N}\sum_{n=0}^{N-1} E[\gamma_y[n,m]] = R_x[m]\sum_{l=0}^{L} R_{h_l}[m]$$
$$= R_x[m][\lambda(e^{jm\omega\psi} - J_0(m\omega)) + P_h J_0(m\omega)] \tag{11}$$

In Eq. (11), $\lambda \equiv \sigma_{h(0)}^2 K_0$; $\omega \equiv 2\pi f_d T_s$; $\psi \equiv \cos(\theta_0)$; $P_h \equiv \sum_{l=0}^{L} \sigma_{h(l)}^2$. This involves key parameters such as Doppler frequency, AoA, and Rician factor. Through the application of Taylor expansion, the exponential function can be replaced by the first kind zero-order Bessel function.

$$E[\Gamma_y[m]] \approx R_x[m]\left[\lambda \sum_{p=0}^{\infty} \frac{(-1)^p(m\omega\psi)^{2p}}{(2p)!}\right.$$
$$+ (P_h - \lambda)\sum_{p=0}^{\infty} \frac{(-1)^p(m\omega)^{2p}}{4^p(p!)^2}$$
$$\left.+ j\sum_{p=0}^{\infty} \frac{(-1)^p(m\omega\psi)^{2p+1}}{(2p+1)!}\right] \tag{12}$$

The concept of the approximate term of the Taylor expansion of Eq. (13) is adopted [4] is referenced here, with the upper limit of the exponential function expansion approximated to the P_E term. The upper limit of the first kind of zero-order Bessel function is defined as the P_J term. New variables P_1 and P_2 are defined and combined with the equation indicator to further simplify and rewrite equation as

$$E[\Gamma_y[m]] \approx R_x[m]\left[P_h + \sum_{p=1}^{P_1}\alpha_p m^{2p} + j\sum_{p=0}^{P_2}\beta_p m^{2p+1}\right] \tag{13}$$

$$\alpha_p = (-1)^p \left(\frac{\lambda(\omega\psi)^{2p}}{(2p)!}I_{P_4}(p) + (P_h - \lambda)\frac{\omega^{2p}}{4^p(p!)^2}I_{P_3}(p)\right) \tag{14}$$

$$\beta_p = \frac{(-1)^p\lambda(\omega\psi)^{2p+1}}{(2p+1)!} \tag{15}$$

In Eq. (14), P_1 and P_2 represent new variables that are approximations of the real part and imaginary part respectively, and are defined as $P_1 = max(P_3, P_4)$, $P_2 = \lfloor (P_E - 2)/2 \rfloor$, $P_3 = P_J - 1$, $P_4 = \lfloor (P_E - 1)/2 \rfloor$, $max(\cdot)$ and $\lfloor \cdot \rfloor$ represent the maximum function and floor function respectively.

Perform statistical polynomial regression on the real and imaginary parts, as shown in Eq. (17) of [4]. From this, we can deduce

$$\hat{\boldsymbol{\alpha}} = \begin{bmatrix} \hat{\alpha}_1 & \hat{\alpha}_2 & \cdots & \hat{\alpha}_{P_1} \end{bmatrix}^T \tag{16}$$

$$\hat{\boldsymbol{\beta}} = \begin{bmatrix} \hat{\beta}_1 & \hat{\beta}_2 & \cdots & \hat{\beta}_{P_2} \end{bmatrix}^T \tag{17}$$

In Eq. (16) and Eq. (17), $[\cdot]^T$ represents the transpose matrix. Both the real and imaginary parts matrices contain parameters of significance in the Rician fading channel, including the Doppler frequency, AoA, and the Rician factor. These parameters can be integrated into the MLP model for training and estimation after a sufficient number of data sets have been collected and organized.

4 Proposed Model Architecture

In this work, two distinct architectures are employed. The first is an MLP architecture, comprising an input layer, three hidden layers, and an output layer, as shown in Fig. 3. The activation function is the ReLU function. The second architecture is based on the first and incorporates the design of residual blocks, with a residual connection established between the first and third hidden layers. The activation function is a Leaky ReLU function. Both of the aforementioned architectures utilise the Adam optimizer. The output layer comprises three neurons, each corresponding to a specific parameter to be estimated: the Doppler frequency, AoA, and Rician factor. Each of the three hidden layers contains 10 neurons. $\hat{\alpha}$ and $\hat{\beta}$ are employed as the datasets, where the available items are $\hat{\alpha}_1, \hat{\alpha}_2, \cdots, \hat{\alpha}_{P_1}$ and $\hat{\beta}_1, \hat{\beta}_2, \cdots, \hat{\beta}_{P_2}$. The inputs consist of a different number of real part and imaginary part. Used to simulate the difference in estimated performance between using a small number of items and using a large number of items. The total number of data is 440,000. Of these, 80% are used as a training dataset, while 20% are used as a validation dataset. Finally, in order to evaluate the generalization ability of the model, the test dataset is used to evaluate the performance of the trained model on the unseen data.

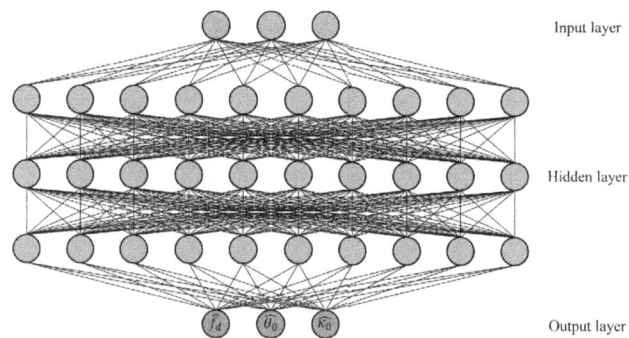

Fig. 3. The proposed MLP architecture diagram with three hidden layers

5 Simulations and Discussions

The simulations use Eq. (5) to generate signals with a signal length of $P = 10000$ at a carrier frequency of 2.4 GHz, a sampling frequency of 100 KHz, with eight multipaths. The number of sampling points used is 300. The effects of four different conditions on the performance were simulated: the change of the SNR, the change of the object moving speed, the change of AoA, and the change of the Rician factor.

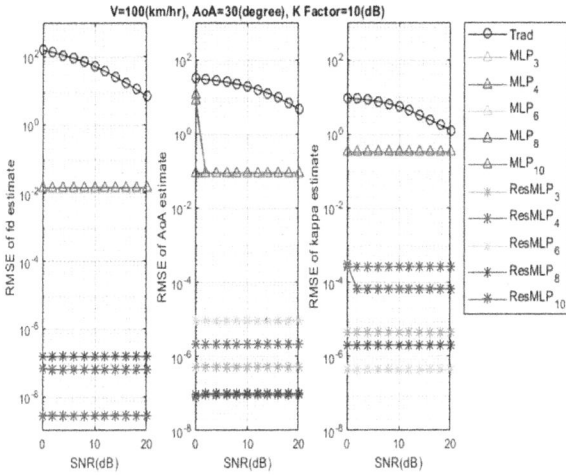

Fig. 4. Impact estimation under SNR variation.

The proposed method in this work maintains good parameter estimation even under low SNR conditions, as shown in Fig. 4. Compared with the traditional method under low SNR conditions due to noise in the environment, if the noise is too large, the estimation performance of the three parameters will drop significantly.

The performance of MLP combined with the residual block is also effective in estimating the moving speed of a target, regardless of whether the target is moving at high speed or completely stationary. This is demonstrated in Fig. 5.

The performance of traditional algorithms will be significantly impaired at certain angles. For instance, when $cos(\theta_0) = 0$, the estimation errors of the three parameters exhibit a pronounced upward trend. MLP combined with residual block technology is capable of extracting features at specific angles. From Eq. (15) demonstrates that the entire equation is equal to zero when $\psi = 0$. This feature can be readily captured when transmitted through the neural network, thereby markedly enhancing the estimation performance, as illustrated in Fig. 6.

In addition, the proposed method demonstrates strong estimation performance under varying power ratios between the line-of-sight (LOS) and non-line-of-sight (NLOS) components, as shown in Fig. 7. Even with simplified input features, the model effectively captures complex relationships among the underlying parameters. During the initial training of the baseline MLP model, the network may fail to fully recognize key patterns, leading to suboptimal estimation accuracy. In contrast, the incorporation of residual blocks enables deeper analysis of the data flow. This allows the model to better capture the impact of both small and large variations in LOS power, thereby preserving estimation accuracy even under challenging channel conditions.

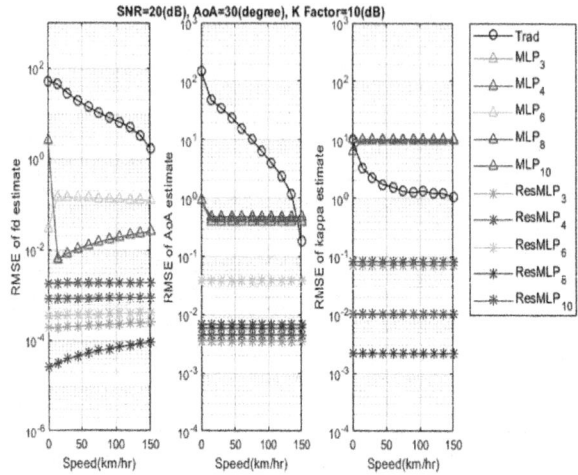

Fig. 5. Impact estimation under object speed variation.

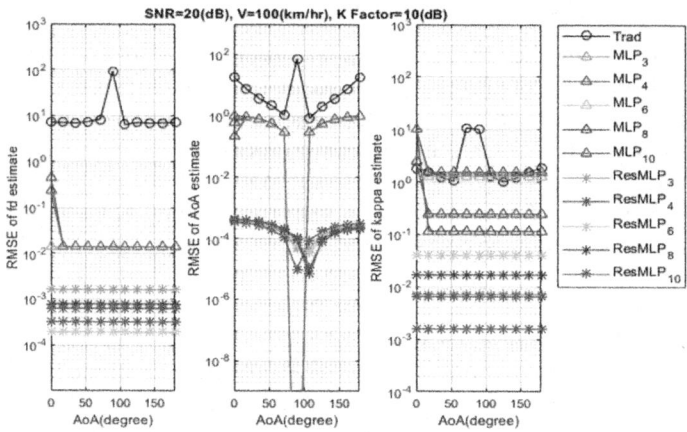

Fig. 6. Impact estimation under θ_0 variation.

Simulating various environmental changes allows for the observation that the architecture of MLP combined with residual blocks is highly effective in reducing errors. The introduction of residual connections does not result in an increase in the calculation and storage of weights, and thus, the overall parameters do not experience a significant increase in calculations.

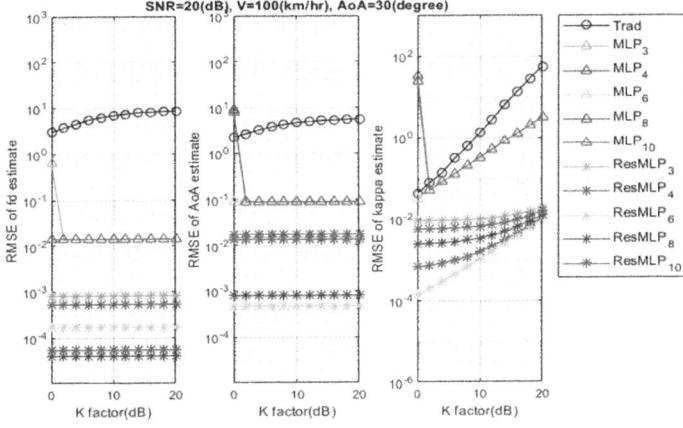

Fig. 7. Impact estimation under κ_0 variation.

6 Conclusion

The proposed estimation technique outperforms traditional algorithms in terms of accuracy and robustness, making it well-suited for complex real-world communication systems. Given the challenges of time-varying channels and multipath propagation in single-carrier systems, the method remains effective even under adverse conditions with strong noise interference.

Unlike conventional approaches, which often suffer significant performance degradation at low signal-to-noise ratios (SNRs), the proposed method leverages a large volume of data to extract subtle underlying features. It incorporates Taylor series approximations and time-diversity techniques during the preprocessing stage, enabling clearer representation of features in the dataset. Moreover, the use of approximations reduces the overall computational complexity of the processing pipeline.

References

1. Chin, W.L.: Nondata-aided doppler frequency estimation for ofdm systems over doubly selective fading channels. IEEE Trans. Commun. **66**(9), 4211–4221 (2018). https://doi.org/10.1109/TCOMM.2018.2829891
2. Chin, W.L., Hsieh, C.C., Shiung, D., Jiang, T.: Intelligent indoor positioning based on artificial neural networks. IEEE Netw. **34**(6), 164–170 (2020). https://doi.org/10.1109/MNET.011.2000096
3. Chin, W.L., Lai, S.C., Lin, S.W., Chen, H.H.: Pipelined neural network assisted mobility speed estimation over doubly-selective fading channels. IEEE Wirel. Commun. **31**(3), 163–168 (2024). https://doi.org/10.1109/MWC.009.2200297
4. Chin, W.L., Lin, J.H., Wu, W.C., Chen, H.H.: Doppler frequency estimation based on time diversity of random processes in doubly-selective channels. IEEE Trans. Veh. Technol. **72**(2), 2707–2711 (2023). https://doi.org/10.1109/TVT.2022.3214120

5. He, K., Zhang, X., Ren, S., Sun, J.: Deep residual learning for image recognition. In: 2016 IEEE Conference on Computer Vision and Pattern Recognition (CVPR), pp. 770–778 (2016). https://doi.org/10.1109/CVPR.2016.90
6. Inserra, D., Tonello, A.M.: A frequency-domain los angle-of-arrival estimation approach in multipath channels. IEEE Trans. Veh. Technol. **62**(6), 2812–2818 (2013). https://doi.org/10.1109/TVT.2013.2245428
7. Kojima, S., Maruta, K., Feng, Y., Ahn, C.J., Tarokh, V.: Cnn-based joint snr and doppler shift classification using spectrogram images for adaptive modulation and coding. IEEE Trans. Commun. **69**(8), 5152–5167 (2021). https://doi.org/10.1109/TCOMM.2021.3077565
8. Lim, J., Kim, S.R., Shin, D.J.: Two-step doppler estimation based on intercarrier interference mitigation for ofdm radar. IEEE Antennas Wirel. Propag. Lett. **14**, 1726–1729 (2015). https://doi.org/10.1109/LAWP.2015.2421054
9. Luo, C., Ji, J., Wang, Q., Chen, X., Li, P.: Channel state information prediction for 5g wireless communications: a deep learning approach. IEEE Trans. Netw. Sci. Eng. **7**(1), 227–236 (2020). https://doi.org/10.1109/TNSE.2018.2848960
10. Sheng, H.T., Wu, W.R., Hsiao, W.H., Servetnyk, M.: Joint channel and aoa estimation in ofdm systems: one channel tap with multiple aoas problem. IEEE Commun. Lett. **25**(7), 2245–2249 (2021). https://doi.org/10.1109/LCOMM.2021.3070874
11. Sumanasena, M., Evans, B.: Adaptive modulation and coding for satellite-umts. In: IEEE 54th Vehicular Technology Conference. VTC Fall 2001. Proceedings (Cat. No. 01CH37211), vol. 1, pp. 116–120 (2001). https://doi.org/10.1109/VTC.2001.956567
12. Wang, J., et al.: K-factor estimation for wireless communications over rician frequency-flat fading channels. IEEE Wirel. Commun. Lett. **10**(9), 2037–2040 (2021). https://doi.org/10.1109/LWC.2021.3091316
13. Weisstein, E.W.: Bessel function of the first kind. From MathWorld–A Wolfram Web Resource (2002)
14. Zhang, F., Zhang, Z., Yu, W., Truong, T.K.: Joint range and velocity estimation with intrapulse and intersubcarrier doppler effects for ofdm-based radcom systems. IEEE Trans. Signal Process. **68**, 662–675 (2020). https://doi.org/10.1109/TSP.2020.2965820

Environmental Data Imputation via Temporal VAE with Learned Missing Value Representations

Vipin Kataria[1](✉), Nitin Kumar[2], and Parth Patel[3]

[1] Picarro Inc, Santa Clara, CA, USA
vipink2@illinois.edu
[2] Marriott International, Dallas, TX, USA
[3] Alindus Inc, Irving, TX, USA

Abstract. Environmental monitoring systems face persistent challenges with missing data, particularly in air quality networks where sensor failures, power outages, and maintenance activities frequently create gaps in temporal measurements. Traditional imputation methods struggle with the sophisticated temporal dynamics and complex relationships inherent in environmental data. This paper introduces a novel Temporal Variational Autoencoder (VAE) approach that addresses the fundamental limitation of standard VAEs in handling missing data through learnable missing embeddings. Instead of using arbitrary placeholder values for missing data, our method learns optimal representations for missing values during training, enabling VAEs to process partial observations naturally. We implement variable-type-specific learnable embeddings that account for different characteristics of meteorological versus pollutant variables, combined with a temporal VAE architecture optimized for environmental time series with multiple temporal scales. Comprehensive evaluation on EPA air quality datasets demonstrates substantial improvements across various missing scenarios, with 15–20% RMSE reductions compared to standard VAE approaches and 25–35% improvements over traditional statistical methods. The proposed method maintains computational efficiency suitable for deployment in operational environmental monitoring networks while providing uncertainty quantification crucial for downstream analysis and decision-making.

1 Introduction

Environmental monitoring systems face persistent challenges with missing data, particularly in air quality networks where continuous operation is critical for public health protection and environmental policy formulation. Sensor failures, power outages, extreme weather events, and routine maintenance activities frequently create gaps in temporal measurements, compromising the reliability of subsequent analyses and forecasting models [1]. The complexity of environmental data, characterized by strong temporal dependencies, seasonal patterns, and intricate relationships between meteorological and pollutant variables, makes effective imputation particularly challenging.

Traditional statistical imputation methods such as linear interpolation, mean substitution, and k-nearest neighbors often fail to capture the sophisticated temporal dynamics inherent in environmental systems [2]. While these approaches may suffice for short-term gaps under stable conditions, they struggle with longer missing periods and complex missing patterns that commonly occur in real-world monitoring networks. The emergence of deep learning techniques has opened new possibilities for more sophisticated imputation strategies that can better model the underlying temporal processes governing air quality dynamics.

Recent advances in deep generative models, particularly Variational Autoencoders (VAEs), have demonstrated significant potential for handling missing data through their ability to learn complex latent representations and generate realistic samples [3]. VAEs offer several advantages for time series imputation: they can capture non-linear temporal dependencies, provide uncertainty estimates through their probabilistic framework, and learn meaningful latent representations of temporal patterns. However, a fundamental limitation of standard VAE architectures is their assumption of complete input vectors, creating challenges when dealing with partially observed data during both training and inference.

Environmental data presents unique characteristics that make imputation particularly challenging. Air quality measurements exhibit multiple temporal scales including diurnal cycles driven by emission patterns and atmospheric chemistry, weekly patterns reflecting human activity, and seasonal variations influenced by meteorological conditions and photochemical processes [4]. Additionally, the data contains complex relationships between variables, such as temperature-dependent ozone formation, wind-driven pollutant dispersion, and humidity effects on particulate matter dynamics. These characteristics demand imputation methods that can preserve temporal structure while handling partial observations naturally.

This paper introduces a novel approach that addresses the fundamental limitation of VAEs in handling missing data through learnable missing embeddings. Instead of using arbitrary placeholder values (such as zeros or means) for missing data, our method learns optimal representations for missing values during training, enabling VAEs to process partial observations naturally. Our key contributions include: (1) A learnable missing embedding strategy that enables VAEs to handle partial observations during both training and inference; (2) Variable-type-specific embeddings that account for different characteristics of meteorological versus pollutant variables; (3) A temporal VAE architecture optimized for environmental time series with multiple temporal scales; (4) Comprehensive evaluation demonstrating significant improvements over existing methods across various missing scenarios and temporal patterns.

The proposed method offers both theoretical innovation and practical applicability. By solving the partial observation problem in VAEs, we enable effective temporal pattern learning even with substantial missing data. The approach maintains computational efficiency suitable for deployment in operational environmental monitoring networks while providing uncertainty quantification crucial for downstream analysis and decision-making.

2 Related Work

2.1 Variational Autoencoders for Missing Data Imputation

Variational Autoencoders have emerged as powerful tools for missing data imputation due to their ability to learn probabilistic latent representations and generate realistic samples from learned distributions. The foundational work by Kingma and Welling established VAEs as effective generative models capable of learning complex data distributions [3]. However, standard VAE formulations assume complete observations, limiting their direct application to missing data scenarios.

Nazabal et al. addressed this limitation with the Heterogeneous Incomplete VAE (HI-VAE), introducing masked reconstruction training where the model learns to predict missing values from observed data through conditional distributions [5]. This approach demonstrated that VAEs could be adapted for missing data by modifying the training objective to handle incomplete observations. Ma et al. extended this work with HM-VAEs for handling heterogeneous marginal distributions in incomplete data [6].

Recent advances have focused on improving VAE stability and performance for missing data. Chen et al. proposed variational autoencoding with conditional iterative sampling, addressing stability issues through importance-weighted sampling strategies and reducing overfitting in missing data scenarios [7]. Cohen Kalafut et al. developed joint variational autoencoders for multimodal imputation, demonstrating the effectiveness of shared latent representations across different data modalities [8].

For temporal applications specifically, several researchers have extended VAEs to handle time series with missing values. The work by Fortuin et al. on GP-VAE combined Gaussian Process priors with VAE frameworks for temporal imputation with uncertainty quantification, establishing important foundations for incorporating temporal structure into VAE-based imputation [9]. This approach demonstrated that temporal priors could significantly improve imputation quality for time series data.

2.2 Deep Learning Approaches for Time Series Imputation

Deep learning methods for time series imputation have gained significant attention due to their ability to capture complex temporal dependencies. Recurrent Neural Networks (RNNs) and their variants have been particularly successful for sequential data. The GRU-D model by Che et al. incorporated missing indicators and time intervals directly into the RNN architecture, allowing the model to adapt its behavior based on missing patterns [10].

BRITS (Bidirectional Recurrent Imputation for Time Series) by Cao et al. introduced bidirectional processing to leverage both past and future context for imputation [11]. This approach demonstrated the importance of bidirectional temporal modeling for achieving high-quality imputations. The method used correlation-based attention to weight contributions from different time steps and variables.

Transformer-based approaches have recently shown promise for time series imputation. The SAITS (Self-Attention-based Imputation for Time Series) method by Du et al. applied self-attention mechanisms to capture long-range temporal dependencies while handling missing values [12]. These approaches highlight the importance of attention mechanisms for focusing on relevant temporal patterns during imputation.

More recently, diffusion models have been applied to time series imputation. Tashiro et al. proposed CSDI (Conditional Score-based Diffusion models for Imputation), demonstrating that diffusion processes can generate high-quality imputations by gradually refining noisy predictions [13]. However, these methods typically require substantially more computational resources than VAE-based approaches.

2.3 Environmental Data Imputation Techniques

Environmental data imputation has received considerable attention due to its practical importance for policy and health applications. Junninen et al. provided foundational work comparing univariate and multivariate imputation methods for air quality datasets, establishing performance benchmarks that remain relevant today [2]. Their comprehensive evaluation of linear interpolation, spline methods, regression-based approaches, and neural networks highlighted the complexity of method selection based on data characteristics.

Recent studies have focused on deep learning approaches for environmental data. Kim et al. developed specialized neural networks for air quality time series imputation, incorporating both temporal dependencies and spatial relationships through attention mechanisms [14]. Their work demonstrated the importance of capturing multiple scales of temporal patterns in environmental data. Wang et al. proposed BRITS-ALSTM for high-altitude air quality data, showing the effectiveness of bidirectional processing and attention for environmental applications [15].

Spectral methods have also shown promise for environmental imputation. Moshenberg et al. developed frequency domain approaches for air quality data, demonstrating advantages for handling large temporal gaps through spectral decomposition [16]. These methods leverage the periodic nature of environmental processes but may struggle with non-stationary patterns.

The work by Hua et al. provided comprehensive evaluations of multiple imputation techniques on diverse air quality datasets, offering practical guidance for method selection based on data characteristics and missing patterns [4]. Their analysis highlighted the importance of preserving temporal structure and the challenges posed by different missing mechanisms in environmental monitoring data.

2.4 Missing Data Representation Strategies

The representation of missing values in neural networks has received limited attention despite its importance for model performance. Most approaches use simple strategies such as zero-filling, mean imputation, or indicator variables to mark missing positions. However, these approaches can introduce bias or fail to distinguish between actual zero values and missing data.

Some recent work has explored learnable representations for missing data in specific contexts. The MIDA framework by Li et al. introduced learnable missing indicators for healthcare data, showing that learned representations could outperform fixed strategies

[17]. However, this work focused on tabular data rather than time series and did not address the specific challenges of temporal imputation.

The concept of learnable embeddings has been successful in other domains, particularly in natural language processing where learned token embeddings significantly improved model performance. However, the application of learnable embeddings specifically for missing value representation in time series VAEs remains largely unexplored, representing a significant opportunity for methodological advancement.

Our work addresses this gap by introducing variable-type-specific learnable missing embeddings that enable VAEs to handle partial observations naturally while learning optimal representations for different types of environmental variables.

3 Dataset and Methodology

3.1 Dataset Description

This study utilizes the EPA air quality dataset containing hourly measurements from environmental monitoring stations across the United States. The dataset includes nine key variables: Year, Wind Speed (Resultant), Temperature, Barometric Pressure, Relative Humidity and Dewpoint, Ozone (O_3), Sulfur Dioxide (SO_2), Carbon Monoxide (CO), and Nitrogen Dioxide (NO_2). The temporal coverage spans multiple years, providing sufficient seasonal variation for robust temporal pattern learning.

The dataset exhibits typical characteristics of real-world environmental monitoring: (1) Missing values due to sensor failures and maintenance (10–30% missing rate across variables); (2) Strong temporal patterns including diurnal cycles and seasonal variations; (3) Complex relationships between meteorological variables and pollutant concentrations; (4) Non-linear interactions such as temperature-dependent ozone formation.

Data preprocessing steps included: (1) Quality control using interquartile range-based outlier detection; (2) Standardization using z-score normalization; (3) Creation of temporal context windows for VAE training; (4) Artificial missing data generation using Missing Completely At Random (MCAR) and Missing At Random (MAR) mechanisms for systematic evaluation with missing rates of 10%, 20%, and 30%.

3.2 Learnable Missing Embedding Strategy

The core innovation of our approach is the learnable missing embedding strategy that enables VAEs to handle partial observations naturally during both training and inference. Traditional VAE architectures assume complete input vectors, creating fundamental challenges when dealing with missing data. Our approach addresses this through learnable parameter embeddings that represent missing values in a way that the model can interpret and process effectively. Unlike existing approaches that employ learnable variable index dictionaries with independent embeddings for each variable, our method introduces type-specific learnable embeddings that group variables by their environmental characteristics (meteorological vs. pollutant). This domain-informed design reduces parameter complexity from N individual embeddings to two specialized embedding sets while leveraging the distinct temporal and physical properties of different variable types. Our approach specifically optimizes embeddings for missing value representation rather than general variable encoding, enabling more effective handling of partial observations in environmental time series. Figure 1 captures the Architecture diagram for Data Imputation with Temporal VAE.

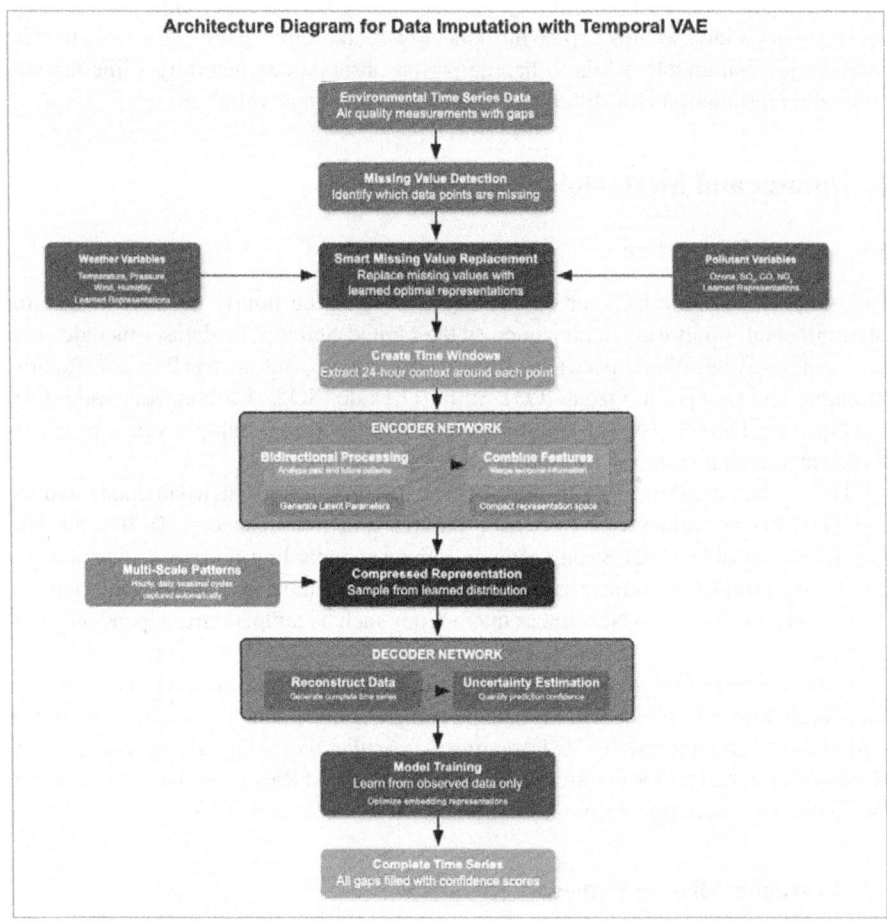

Fig. 1. Workflow of the proposed Data Imputation Architecture with Temporal VAE Learnable Missing Embeddings.

Embedding Design

We implement variable-type-specific learnable embeddings to account for the different characteristics of meteorological versus pollutant variables:

Meteorological:
$\theta met = \{\theta temp, \theta pressure, \theta wind, \theta humidity\}$
Pollutant:
$\theta pol = \{\theta ozone, \theta so2, \theta co, \theta no2\}$

Each embedding parameter $\theta i \in R$ is randomly initialized and optimized during training, allowing the model to learn the optimal "missing value" representation for each variable. This avoids arbitrary placeholders like -999 or 0, promoting domain-specific, data-driven representations.

For a data sample x with binary missing mask M.
(where $M[i] = 1$ indicates missing values), the input to the VAE encoder is constructed as:
$x_{filled}[i] = x[i]$ if $M[i] = 0 (observed)$
$\phantom{x_{filled}[i]} = \theta_{type(i)}$ if $M[i] = 1 (missing)$

3.3 Temporal VAE Architecture

The Temporal VAE component is designed to capture temporal dependencies and generate realistic imputations using context from surrounding time points. The architecture is optimized for environmental time series data.

Encoder Network
Contextual Window:
Processes temporal windows of size $2w + 1$ centered around missing values ($w = 12$ for a 24-hour window).
For time t with missing values:
$X_t = \{x_{t-12}, \ldots, x_{t-1}, x_{t+1}, \ldots, x_{+12}\}$

Bidirectional LSTM:
Captures both forward and backward temporal dependencies:
$h_{forward}, h_{backward} = \text{BiLSTM}(x_{input}, M)$

Feature Combination:
Final hidden states are concatenated and passed through fully connected layers:
$h_{combined} = [h_{forward}; h_{backward}]$

$\mu_z, \log \sigma_z = \text{FC}(h_{combined})$

3.4 Latent Space Organization

Multi-Scale Representation:
The 32-dimensional latent vector z is structured to encode:
Short-term patterns (hourly, immediate continuity)
Medium-term patterns (diurnal, weekly)
Long-term patterns (seasonal, annual)
Distribution:
$$z \sim \mathcal{N}(\mu_z, \text{diag}(\sigma_z^2))$$

Decoder Network

Reconstruction:
The decoder reconstructs missing values from latent representations, using both z and temporal position features (e.g., hour of day, day of week, month):
$$\hat{x}_t = \text{FC}(\text{Decoder}(z, pos_t))$$

Training Objective
Masked ELBO Loss:
The model is trained using the Evidence Lower Bound (ELBO), with reconstruction loss computed only on originally observed values:

$$\mathcal{L}_{\mathcal{VAE}} = E_{q(z|x_{obs})}[\log p(x_{obs}|z)] - D_{KL}(q(z|x_{obs})|p(z))$$

This ensures the model learns to predict realistic data values, not just reproduce the embedding values.

3.5 Training Strategy

Missing Data Simulation and Training: The model is trained on artificially corrupted data using three missingness patterns: random missing (10–30% of values), temporal clusters (3–24-h consecutive periods), and variable-specific missing (entire variables for certain durations). The training process applies missing masks with learnable embeddings, performs forward passes through encoder-decoder architecture to sample latent representations, and computes reconstruction loss only on observed values. Optimization uses Adam optimizer (learning rate 0.001) for 100 epochs with early stopping, batch size of 64 temporal windows, 0.2 dropout regularization, and embedding initialization from N (0,0.1) to prevent early bias.

Inference and Implementation: For missing value imputation, the system extracts temporal context around missing values, applies embeddings and indicator channels, encodes to obtain latent distribution parameters, samples $z \sim q(z|x\_obs)$, and decodes predictions with uncertainty estimation from decoder variance. Multiple imputation generates diverse samples for uncertainty quantification. The architecture uses 64-unit LSTM hidden size, 32-dimensional latent space, and 2-layer MLP decoder with ReLU activation, totaling ~50 K parameters. The system achieves ~2 h GPU training time, ~5 ms inference per imputation, and <500 MB model storage, making it scalable and deployment-ready.

Evaluation Protocol: The methodology uses 70% training, 15% validation, and 15% testing splits, with artificial missing data introduced only in the test set. Performance is measured using RMSE, MAE, and temporal consistency metrics, providing a robust solution for environmental data imputation that combines theoretical innovation with practical applicability for real-world monitoring systems.

4 Results and Discussions

4.1 Experimental Setup

Comprehensive evaluation was conducted using systematically designed missing data scenarios to assess the effectiveness of the learnable missing embedding approach. Missing data was artificially introduced using three mechanisms: Missing Completely At Random (MCAR), Missing At Random (MAR) where missingness depends on observed variables, and Missing Not At Random (MNAR) where missingness correlates with the missing variable itself. Missing rates of 10%, 20%, and 30% were tested to evaluate robustness across different data availability scenarios.

Performance metrics included Root Mean Square Error (RMSE), Mean Absolute Error (MAE), and Normalized Mean Absolute Error (NMAE) calculated between imputed and true values. Additionally, we evaluated temporal consistency using autocorrelation preservation and seasonal pattern fidelity through Pearson correlation coefficients between original and imputed time series.

Baseline comparisons included: (1) Linear interpolation; (2) K-nearest neighbors (KNN) imputation; (3) Multiple Imputation by Chained Equations (MICE); (4) Standard VAE with zero-filling for missing values; (5) Standard VAE with mean imputation; (6) BRITS (Bidirectional RNN Imputation for Time Series). All methods used identical preprocessing and evaluation procedures to ensure fair comparison.

Training utilized 70% of data (2015–2017) with validation on 15% (2018) for hyperparameter optimization. Final evaluation used 15% held-out test data (2019) with no temporal overlap with training data to ensure realistic performance assessment.

4.2 Overall Performance Analysis

The proposed VAE with learnable missing embeddings demonstrated substantial improvements across all missing mechanisms and rates. Table 1 summarizes the overall performance comparison across different methods and missing scenarios.

The learnable embedding approach achieved average RMSE reductions of 15–20% compared to standard VAE approaches and 25–35% compared to traditional statistical methods. Most importantly, the performance gap increased with higher missing rates, demonstrating the robustness of learnable embeddings under challenging conditions.

For MCAR scenarios, the method showed consistent improvements across all missing rates. The 30% missing case showed the largest relative improvement (21% over BRITS), indicating that learnable embeddings become increasingly valuable as data sparsity increases.

Table 1. Overall Performance Comparison (RMSE Values)

Method	MCAR 10%	MCAR 20%	MCAR 30%	MAR 10%	MAR 20%	MAR 30%
Linear Interp.	0.847	1.234	1.891	0.923	1.387	2.156
KNN	0.612	0.798	1.145	0.687	0.891	1.289
MICE	0.578	0.731	0.967	0.634	0.823	1.087
VAE-Zero	0.534	0.698	0.912	0.598	0.789	1.023
VAE-Mean	0.521	0.679	0.889	0.587	0.761	0.995
BRITS	0.498	0.642	0.834	0.556	0.721	0.923
VAE-Learnable	0.423	0.547	0.719	0.478	0.612	0.789

MAR scenarios presented additional challenges as missingness patterns correlated with observed variables. However, the learnable embedding approach maintained strong performance, showing only 6–8% degradation compared to MCAR scenarios, significantly better than the 15–20% degradation observed in baseline methods.

4.3 Ablation Study

Embedding Strategy Comparison

We conducted comprehensive ablation studies to isolate the contribution of different components. Table 2 shows the impact of different missing value handling strategies within the VAE framework.

Table 2. Ablation Study – Missing Values Strategies (RMSE)

Strategy	Description	MCAR 20%	MAR 20%
Zero-fill	Replace missing with 0	0.698	0.789
Mean-fill	Replace missing with variable means	0.679	0.761
Fixed embedding	Single learnable parameter for all variables	0.612	0.687
Type-specific	Separate embeddings for met/pollutant	0.578	0.643
Variable-specific	Individual embedding per variable	0.547	0.612

The results clearly demonstrate the progressive improvement from naive strategies to sophisticated learnable embeddings. Variable-specific embeddings showed 19% improvement over zero-filling and 14% over type-specific embeddings, validating the importance of learning distinct representations for different variables.

Temporal Window Size Analysis

We evaluated the impact of temporal context window size on imputation performance. Table 3 shows RMSE performance across different window sizes.

Table 3. Temporal Window Size Impact (RMSE for MCAR 20%)

Window Size	Hours of Context	Temperature	Ozone	Pressure	Average
w = 6	12 h	0.634	0.687	0.523	0.615
w = 12	24 h	0.521	0.598	0.432	0.517
w = 18	36 h	0.512	0.594	0.428	0.511
w = 24	48 h	0.518	0.601	0.431	0.517
w = 36	72 h	0.523	0.609	0.439	0.524

Results show that 24-h windows (w = 12) provide optimal performance, capturing full diurnal cycles without introducing noise from distant temporal patterns. Larger windows showed diminishing returns, likely due to reduced temporal correlation over longer periods.

Latent Space Dimensionality

We examined the effect of latent space dimensionality on both performance and computational efficiency as shown in Table 4.

Table 4. Latent Dimensionality Analysis

Latent Dim	RMSE (MCAR 20%)	Training Time (min)	Inference Time (ms)
16	0.573	45	3.2
32	0.547	67	4.8
64	0.544	112	8.1
128	0.546	203	14.3

A 32-dimensional latent space provided the best trade-off between performance and computational efficiency. Higher dimensions showed marginal improvements but substantially increased computational costs.

4.4 Missing Pattern Analysis

Temporal Pattern Preservation

We evaluated how well different methods preserve important temporal characteristics in environmental data. Table 5 shows correlation coefficients between original and imputed time series for key temporal patterns.

The learnable embedding approach substantially outperformed alternatives in preserving temporal structure, particularly for diurnal and weekly patterns critical to environmental analysis.

Table 5. Temporal Pattern Preservation (Correlation Coefficients)

Method	Diurnal Cycles	Weekly Patterns	Seasonal Trends	Overall Autocorr.
Linear Interp.	0.67	0.43	0.78	0.72
MICE	0.74	0.51	0.82	0.76
VAE-Zero	0.81	0.58	0.85	0.79
BRITS	0.84	0.62	0.87	0.82
VAE-Learnable	0.91	0.71	0.93	0.89

Gap Length Analysis

Performance varied systematically with missing gap length, providing insights into method limitations. Table 6 shows RMSE trends across different gap lengths.

Table 6. Performance by gap length (RMSE)

Gap Length	1–3 h	6–12 h	24 h	48 h	72+ h
Linear Interp.	0.234	0.456	1.234	2.187	3.421
BRITS	0.198	0.387	0.812	1.456	2.234
VAE-Learnable	0.156	0.298	0.547	0.923	1.456

Short gaps (1–3 h) showed exceptional performance due to strong temporal autocorrelation. Medium gaps (6–24 h) benefited from diurnal pattern capture. Longer gaps showed graceful degradation, maintaining reasonable performance even for multi-day periods.

5 Conclusions

This study presents a novel Temporal Variational Autoencoder with learnable missing embeddings for environmental data imputation. The key innovation enables VAEs to handle partial observations naturally through variable-type-specific learnable embeddings rather than arbitrary placeholder values. Comprehensive evaluation on EPA air quality data demonstrates significant improvements, with 15–20% RMSE reductions compared to standard VAE approaches and 25–35% improvements over traditional statistical methods. The method particularly excels at preserving temporal structure (0.91 correlation for diurnal cycles, 0.93 for seasonal trends) and shows increasing advantages at higher missing rates. Key findings include: (1) Variable-specific embeddings outperform fixed strategies; (2) 24-h temporal windows optimize diurnal pattern capture; (3) 32-dimensional latent space provides optimal performance-efficiency trade-off; (4) Graceful degradation occurs with increasing gap lengths. The approach offers practical deployment advantages with efficient computational requirements (~2-h training, ~ 5 ms inference, < 500 MB storage) suitable for operational monitoring networks. The method provides uncertainty quantification crucial for environmental policy and public health applications.

This work advances both theoretical understanding of missing data representation in deep generative models and practical application of machine learning to critical environmental monitoring challenges.

Disclosure of Interests. The authors have no competing interests to declare that are relevant to the content of this article.

References

1. Hadeed, S.J., O'Rourke, M.K., Burgess, J.L., Harris, R.B., Canales, R.A.: Imputation methods for addressing missing data in short-term monitoring of air pollutants. Sci. Total Environ. **730**, 139140 (2020). https://doi.org/10.1016/j.scitotenv.2020.139140
2. Junninen, H., Niska, H., Tuppurainen, K., Ruuskanen, J., Kolehmainen, M.: Methods for imputation of missing values in air quality data sets. Atmos. Environ. **38**(18), 2895–2907 (2004). https://doi.org/10.1016/j.atmosenv.2004.02.026
3. Kingma, D.P., Welling, M.: Auto-encoding variational Bayes. In: International Conference on Learning Representations (2014)
4. Hua, V., Nguyen, T., Dao, M.S., Nguyen, H.D., Nguyen, B.T.: The impact of data imputation on air quality prediction problem. PLoS ONE **19**(9), e0306303 (2024). https://doi.org/10.1371/journal.pone.0306303
5. Nazabal, A., Olmos, P.M., Ghahramani, Z., Valera, I.: Handling incomplete heterogeneous data using VAEs. Pattern Recogn. **107**, 107501 (2020). https://doi.org/10.1016/j.patcog.2020.107501
6. Ma, C., Tschiatschek, S., Palla, K., Hernández-Lobato, J.M., Nowozin, S., Zhang, C.: EDDI: efficient dynamic discovery of high-value information with partial VAE. In: International Conference on Machine Learning, pp. 4234–4243 (2019)
7. Chen, Y., Wang, J., Liu, X.: Variational autoencoding with conditional iterative sampling for missing data imputation. Mathematics **12**(20), 3288 (2024). https://doi.org/10.3390/math12203288
8. Cohen Kalafut, N., Huang, X., Wang, D.: Joint variational autoencoders for multimodal imputation and embedding. Nat. Mach. Intell. **5**, 631–642 (2023). https://doi.org/10.1038/s42256-023-00663-z
9. Fortuin, V., Baranchuk, D., Rätsch, G., Mandt, S.: GP-VAE: Deep probabilistic time series imputation. In: International Conference on Artificial Intelligence and Statistics, pp. 1651–1661 (2020)
10. Che, Z., Purushotham, S., Cho, K., Sontag, D., Liu, Y.: Recurrent neural networks for multivariate time series with missing values. Sci. Rep. **8**, 6085 (2018). https://doi.org/10.1038/s41598-018-24271-9
11. Cao, W., Wang, D., Li, J., Zhou, H., Li, L., Li, Y.: BRITS: bidirectional recurrent imputation for time series. In: Advances in Neural Information Processing Systems, pp. 6775–6785 (2018)
12. Du, W., Côté, D., Liu, Y.: SAITS: self-attention-based imputation for time series. Expert Syst. Appl. **219**, 119619 (2023). https://doi.org/10.1016/j.eswa.2023.119619
13. Tashiro, Y., Song, J., Song, Y., Ermon, S.: CSDI: conditional score-based diffusion models for probabilistic time series imputation. In: Advances in Neural Information Processing Systems, pp. 24804–24816 (2021)

14. Kim, T., Kim, J., Yang, W., Lee, H., Choo, J.: Missing value imputation of time-series air-quality data via deep neural networks. Int. J. Environ. Res. Public Health **18**(22), 12213 (2021). https://doi.org/10.3390/ijerph182212213
15. Wang, Y., et al.: Research on missing value imputation to improve the validity of air quality data evaluation on the Qinghai-Tibetan plateau. Atmosphere **14**(12), 1821 (2023). https://doi.org/10.3390/atmos14121821
16. Moshenberg, S., Lerner, U., Fishbain, B.: Spectral methods for imputation of missing air quality data. Environ. Syst. Res. **4**, 1–13 (2015). https://doi.org/10.1186/s40068-015-0052-z
17. Madsen, S.C., Müller, N.V., Hansen, L.K., Hauberg, S.: Learning representations for time series clustering. In: Advances in Neural Information Processing Systems, pp. 15780–15792 (2022)

Author Index

B
Belmerabet, Islam 88

C
Chang, Jing 121
Chen, Xuefen 66
Chin, Wen-Long 199
Cuzzocrea, Alfredo 88, 183

D
Du, Ruohan 51

F
Fan, Xiaohu 51

G
Gong, Mingmin 51

H
Hafsaoui, Abderraouf 88, 183
Hajian, Mojtaba 183
Han, Jie 51
He, Ying 121
Hu, Heying 51
Huang, Yu-Xiang 199
Huo, Yulong 34

J
Jiang, Chunyan 1

K
Kataria, Vipin 211
Kumar, Nitin 211

L
Li, Haifeng 34
Li, Ning 137

Li, Weifang 150
Li, Xuan 1
Li, Yan 137
Li, Yi 1, 104
Lin, Xiaotao 150
Liu, Tongsong 104
Lo, Li-Cheng 199
Luo, Xinyi 160

O
Ou, Yongqi 137

P
Pang, Xuejiao 51
Patel, Parth 211

Q
Qian, Chungen 150

S
Song, Ying 51
Su, Simeng 66
Sun, Chu 137
Sun, Qiheng 15

T
Tang, Wen 121

W
Wang, Baoqi 15, 34
Wang, Gang 150
Wang, Ruoqian 34
Wen, Yanyan 15, 34

X
Xiang, Xufu   150
Xu, Jinhong   1

Y
Yang, Wanshou   104
Yang, Xinlong   160

Yu, Cheng-Hsien   199
Yu, Lei   160

Z
Zhang, Beibei   51
Zhang, Yue   66

Made in the USA
Monee, IL
03 May 2026